GCSE
ENGINEERING

Steve Cushing

Orders: please contact Bookpoint Ltd, 130 Milton Park, Abingdon, Oxon OX14 4SB. Telephone: (44) 01235 827720. Fax: (44) 01235 400454. Lines are open from 9.00–6.00, Monday to Saturday, with a 24-hour message answering service. You can also order through our website **www.hodderheadline.co.uk**.

British Library Cataloguing in Publication Data
A catalogue record for this title is available from the British Library

ISBN 0 340 81408 X

This edition published 2004

Impression number   10 9 8 7 6 5 4 3 2 1

Year   2007   2006   2005   2004

Cover photo from TEK IMAGE/SCIENCE PHOTO LIBRARY

Typeset by Phoenix Photosetting, Chatham, Kent

Printed in Great Britain for Hodder & Stoughton Educational, a division of Hodder Headline Plc, 338 Euston Road, London NW1 3BH by J.W. Arrowsmith Ltd., Bristol

Acknowledgements

The author would like to thank Nyree Williams for her support and input, and Diana Spencer and James de Winter for assisting with the content.

The publishers would like to thank the following individuals and institutions for permission to reproduce copyright material:

© Action Plus – 2.9d; © Air Max – 3.14, 3.22a; © ANNEBICQUE BERNARD/CORBIS SYGMA – 3.6; © B.D.I. Images – 1.9, 1.11, 1.14, 1.15, p. 28, 2.26, 2.29a, 2.35, 2.46, 2.52b, 2.56, 3.5; © Bill Miles/CORBIS – 2.49; © Brownie Harris/CORBIS – 2.41, 3.1; © Charles O'Rear/CORBIS – 3.16; © Christie's Images/CORBIS – 2.68b; © CORBIS – 2.36, 2.40b, 2.53, 3.4; © Craig Aurness/CORBIS – 3.12; © Dave Bartruff/CORBIS – 2.62; © David J. & Janice L. Frent Collection/CORBIS – 2.9c; Digital image © 1996 CORBIS, Original image courtesy of NASA/CORBIS – 2.18b; © Dyson 2.2b; © Eric Hausman/CORBIS – 2.29b; Floor Standing Pillar Drill courtesy of Jack Sealey Ltd, Bury St Edmunds – 2.52a; © Gary D. Landsman/CORBIS – 1.16; © Hodder Arnold archives – 2.9b, 3.8; © Image Bank/Getty Images – 1.26b; © James Leynse/CORBIS – 2.17; © Jean Heguy/CORBIS – 2.40a; © Jean Miele/CORBIS – 2.21; © Jose Luis Pelaez, Inc./CORBIS – 2.68a; © Kim Sayer/CORBIS – 1.10; © L. Clarke/CORBIS – 3.19; © Lester Lefkowitz/CORBIS – 2.63; © Life File – 1.4, 1.8; © Michael S. Yamashita/CORBIS – 3.20; © NASA/Roger Ressmeyer/CORBIS – 3.9; © Paul Hutley, Eye Ubiquitous/CORBIS – 1.26a; © Paula Weidiger, Edifice/CORBIS – 2.9a; © Peggy & Ronald Barnett/CORBIS – 2.67; © Peter Steiner/CORBIS – 2.23; © Peter Vadnai/CORBIS – 3.3; © PICIMPACT/CORBIS – 1.12; © REUTER RAYMOND/CORBIS SYGMA – 2.57; © REX – 1.5, 1.18; © Roy Morsch/CORBIS – 2.50; © Salvador Dali AKG – 2.39; © Simon Bruty/SI/NewSport/Corbis – 2.19, 3.22b; © Steve Chenn/CORBIS – 3.7; © Still Pictures – 3.21; © Terry W. Eggers/CORBIS – 3.13; © The National Grid Company plc – 2.18a; © The Ronald Grant Archive – 1.13, 3.15; © TopFoto – 3.18; © William Taufic/CORBIS – 2.61.

Fig 2.69 reproduced with kind permission from the Health and Safety Executive.

Figures 1.6 and 1.7 reproduced with kind permission from the British Standards Institution.

Internal artwork by Daedalus Studios and Barking Dog Art.

Every effort has been made to obtain necessary permission with reference to copyright material. The publishers apologise if inadvertently any sources remain unacknowledged and will be glad to make the necessary arrangements at the earliest opportunity.

Contents

GCSE Engineering

Introduction

This book is written for students studying for the Applied GCSE in Engineering. It aims to help and support them in all sections of the course, providing advice on both practical and theoretical areas of the specification.

Applied GCSE courses are intended to be vocational in nature and are double awards. That means that they are worth two GCSEs rather than one. The Applied GCSE in Engineering focuses on the practical aspects of engineering design and production, and the essential information that students need to underpin their work.

The vocational nature of the course means that the emphasis for assessment is on practical work, presented as a portfolio of evidence. The evidence must be related to real-life problems, such as how to package a fragile product so that it gets to the customer in perfect condition. If possible, this evidence should be based on a real organisation that the student can visit and discuss requirements with. If this is not possible, teachers will need to ensure that the student has sufficient case study material to ensure that the task is realistic and challenging.

Visits to engineering establishments should form an important method of studying for this course. No amount of classroom activity can replace the experience gained from seeing real engineering work taking place. It is exciting and interesting to see engineering in action, and visits will provide an element of careers guidance as well as vital background information for the GCSE course. It is hard for any student to imagine the range of career opportunities presented by engineering without actually seeing what goes on in real life. Design and drafting, procurement and production planning, quality control and testing are roles that students might want to consider in the future, as well as practical skill areas such as machining and welding.

This book is divided into three main sections:

Chapter 1 focuses on Unit 1 of the GCSE specification. It aims to help students understand a design brief and draw up a specification. Students need to carry out investigations and this chapter will help them plan their investigations and apply their skills, knowledge and understanding to the engineering problem they are considering. The examples in the book can be used to develop skills and knowledge and provide a starting point for their real-life investigation, so that when they visit a real engineering company they are well prepared for what they will find. The activities provided will offer a practical focus for this important task, and the examples of workshop drawings and design briefs will give an indication of what the student is expected to do.

Chapter 2 explores the manufacturing methods used in engineering, and covers Unit 2 of the GCSE specification. It explains the importance of quality control standards and the skills and techniques needed to meet those standards. Basic skills, tools and techniques are covered to provide a sound foundation for understanding the production of engineering products. The section on health and safety covers the regulations and the way they are applied in practical terms through risk assessments and safety procedures.

Chapter 3 covers Unit 3, the use of new technology in the world of engineering, whether in the area of new materials, manufacturing methods or environmental considerations. As this unit is assessed by an external test, it provides knowledge, revision notes and practice questions. It is important that students carry out their own research into these exciting areas that are constantly changing. Clear notes will make revising for the examination very much easier. Students will also find information for this unit as they build up their portfolio. Making notes as they go along will help build the knowledge needed for the examination and help provide a real understanding of the process under discussion.

Building your portfolio

The practical work that forms such a vital part of this course needs to be supported by a portfolio of evidence, and students need to understand how to present evidence that demonstrates all aspects of the design and production process.

The portfolio should only contain the evidence required by the assessment evidence grid and should be kept separately from class notes. The evidence required for each individual unit should be kept separate and complete.

Work should be presented in an organized way that is easy to follow. Each section should be labelled and each piece of work should have a heading, so that it is clear what it is intended to demonstrate. The sections should follow the assessment grid closely so that it is easy to check that all areas have been covered. Pages should be numbered and a table of contents produced. Students should include their name and centre number on each page, and candidate number if possible. The portfolio should contain drawings and photographs to illustrate the progress of the work and these should be annotated to show exactly how the product was produced and what tools and techniques were involved. Any problems met along the way should be described, together with how they were solved.

Designing Engineering Products

1

CHAPTER AIMS & INTRODUCTION

This chapter will explore:

- manufacturing processes
- design briefs
- the design process
- quality and design criteria
- the importance of planning
- maintenance
- cost and production
- research
- scientific principles
- communicating with drawings
- tolerances
- the use of new technology.

This chapter of the book covers the first part of the design process. It discusses the information that should be in a design brief, and the specifications that should be drawn up from it. Developing a specification into a design takes research, so that is covered too.

There is a lot of information about planning, because it is so very important. However tempting it is to rush in and make something, badly-planned products have a habit of going wrong. A little more thinking time will help to sort out potential difficulties so that production goes much more smoothly.

There is a section on types of drawing – these need practice, but they then become a designer's most valuable tool in defining and presenting a product.

Computers can be used in many stages of the process, from drawing up project schedules to manipulating images or carrying out strength calculations.

1 INTRODUCTION

All engineering is based upon the manufacturing process, which involves an input, a process, an output and feedback.

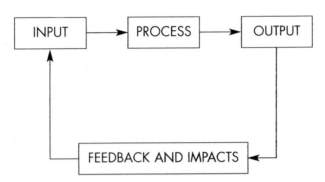

Fig 1.1 A process cycle

Input

The input for any manufacturing system is everything that is needed to start and maintain the process. The input could include money to buy tools, raw materials, machines, buildings and energy to start production. Input also covers the initial idea and the research that takes place to make sure that somebody wants to buy or use the product.

Process

The process is the designing, making and finishing of a product. The process therefore includes all the jobs associated with designing, producing and selling a product. These processes are often grouped into categories:

- design engineering
- production planning
- management
- processing materials
- marketing
- accounting.

Output

The output is everything that exists after, and as a result of, manufacturing. The most obvious output is the product itself, but there are lots of other types of output. Scrap, waste and, sometimes, pollution are the unwanted output of the manufacturing process. Scrap usually includes excess materials, which can sometimes be re-used or recycled. Waste is an output that cannot be recycled and somehow has to be disposed of safely. Pollution is often invisible.

Feedback

Feedback is a method of monitoring and adjusting the input. Feedback is important throughout the manufacturing process. It can take place at lots of different levels; for example, at a micro level, quality control is a feedback. At a higher level, manufacturing processes can be modified as a result of consumer feedback or can be used to reduce adverse effects on the environment.

An Example of a Manufacturing Process

In producing a bike, the manufacturer will research existing bikes and make sure someone wants to buy the product. The manufacturer will raise the necessary money and ensure that they have the right tools, machines and workers. All of these are input.

In making the bike, the manufacturer may use casting, forming, welding, assembling and finishing. It may also study customer needs in order to develop and improve the product. It will produce manufacturing plans, quality control logs and advertising materials. All of these things are processes.

Manufacturing a bike will also produce a number of undesirable waste products e.g. scrap. These are the output, in addition to the bike itself.

The manufacturer will monitor sales and may conduct market research to find out what customers think of the bicycle. The manufacturer will explore any flaws or design errors, will conduct safety checks and may explore environmental issues. All of these things are feedback and may be used to improve the process by changing one or more of the inputs.

True feedback should be based on the full impact of the product and the manufacturing process in terms of the way they affect individuals, society and the environment.

KEY WORDS

Input   the ideas and raw materials that go to create a product

Output   the final product and anything else that is produced such as scrap or waste

Processing   the work that is carried out on the inputs to produce the output

Feedback   checking the outputs and using them to adjust the inputs to make better products

PORTFOLIO NOTES

Once you have a clear idea of what your product needs to do, simple block diagrams of input, output and processing can help to explain your system so it can be clearly understood.

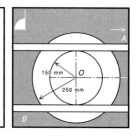

ANALYSING CLIENT DESIGN BRIEFS ⬡2

Before you can even begin to design anything you need to understand what the client wants. You can then look at the design brief and decide what the key features are. The design brief should only explain the problem and not suggest what the possible answers could be. You may be involved with the client in developing or changing the design brief.

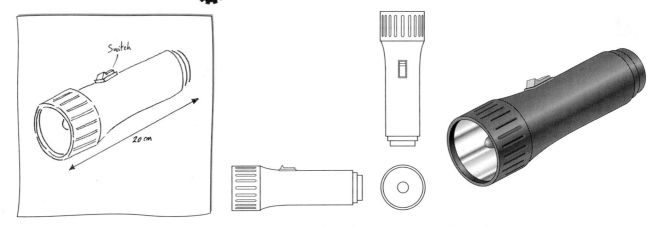

Fig 1.2 A torch – from initial idea to design drawing to finished product

For any engineering product or service some of the key features are:

- quality standards
- styling aesthetics
- performance
- intended markets
- size
- maintenance
- production methods
- production materials
- cost
- regulations
- scale of production.

A product may involve all or only some of these features. We will look at each of these in more detail over the coming pages.

You must not start the design process with a definite idea of what you think will be the best solution. It is important to be open to all ideas even if they seem unlikely at first. If you think widely to begin with, you can narrow your ideas down later. With all these things to consider there will usually be more than one solution that meets the needs of the client (sometimes there will be lots!). Once you have considered all the key features you will have to learn how to evaluate the strengths and weaknesses of each possible solution.

Fig 1.3 Refining your ideas is important

ACTIVITIES

1. State whether the following statements are true or false:

   a. the colour of a finished object must be in the design brief

   b. there will always be one solution to every design problem that is better than all others in every respect

   c. it is important to keep the design brief as clear as possible

   d. the design brief should include suggestions for the final design.

2. Which of the following best describes the design brief for the invention of the DVD?

   a. Design a CD-type disc that can hold full films.

   b. Design a system to replace VHS videotape.

   c. Design a system that can hold more information than either a CD or a VHS videotape.

   d. Find a way to release films that contains extra scenes and information.

3. Try to write a good design brief for the following:

   a. a pencil

   b. a forklift truck

   c. a radio.

KEY WORDS

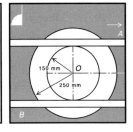

Design brief   statement of what is required from the product

Client   person or company the product is being designed for

PORTFOLIO NOTES

You will need to include a clear design brief that states what the product needs to do without indicating how it might do it.

Key Features

When you start on any design, it is important to see the design process as being made up of many stages. Many of these stages need to be completed before you move on to the next one.

Function

Once you have the design brief you should have a clear idea of what it is that the object or service is supposed to do. This is known as its function. It may be that the object or service needs to provide more than one function and this needs to be considered in your work.

The function of a mobile phone has changed over the last few years. When originally designed, it was expected to allow two-way phone conversations and nothing else. Now the list of functions can include storing phone numbers and names, sending text messages, taking and sending pictures, playing multiple ring tones, playing MP3s, and so on.

Figure 1.4 Early mobile phones were only designed for two-way conversations

Figure 1.5 A state-of-the-art mobile phone even has a TV screen!

ACTIVITIES

1. For each of the following items state their function:

(For example: Object – Light Bulb

Function – To produce light from electricity)

  a. a bicycle

  b. a lawn mower

  c. a match

  d. a pair of glasses.

2. What object would be used to perform each of the following functions?

  a. Store chemical energy to convert into electricity.

  b. Allow visual information from a computer to be seen.

  c. Keep someone's money and credit cards in one place.

  d. Record images that can later be stored on, and seen on, a computer.

KEY WORD

Function   something the product must do

PORTFOLIO NOTES

The design brief should clearly state what is needed, but you should list all the functions that are required. It is then easy to check this list against your design. If you work this way, you will make sure that you do not leave anything out.

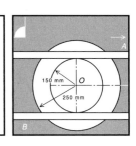

Quality Standards

Quality is usually said to mean whether something is 'fit for its purpose' i.e., will it do what it is supposed to? So the quality standards are the list of requirements that the product or service must satisfy. The client is the one who is most likely to set the standards. Other requirements may come from other places: there may be standards set by the sector and there may also be standards set by legislation.

Some of the main areas considered when looking at quality are as follows.

- The design – does the designed product match the client's needs? It is also important to check that the final product matches the design.

- Reliability – how long will it be before a product fails? Is it easily replaced without further problems?

- Service and maintenance – who is responsible when problems occur? Who is responsible for replacement? These areas are usually covered in the contract for purchase and service.

There is a variety of legislation that covers some of these areas, included in the British Standard BS EN ISO 9001: 2000.

Figs 1.6 and 1.7 Examples of accreditation marks used by the British Standards Institution

If you were designing a swing, it should be able to take the weight of a child. This may be the only requirement from the client. It may be that all makes of swing must be sure that they can take the weight of an adult – this is the standard set by the sector. There may be a law that states that the swing must have a sign or sticker stating the maximum weight it may take – this is a standard set by legislation.

Your brief may be to design a piece of fire safety equipment that is able to put out a small electrical fire (e.g. a fire in a computer or a television set). You choose to design a fire extinguisher.

The client may specify a certain, size shape and colour, but there are laws about the size, shape and colour that fire extinguishers are allowed to be. In this case, you may need to go back to the client to change the brief.

The life span or reliability of a product may also be specified. For example a light bulb may need to last 1000 hours or a swing for five years of regular use.

Quality control techniques are dealt with in more detail later in this book.

Fig 1.8 A swing needs to be able to take the weight of a child

Fig 1.9 There are specific regulations to bear in mind when designing a fire extinguisher

ACTIVITY

Try to list as many quality standards as possible that might apply to each of the following objects:

- a milk bottle

- a bicycle

- a glue stick.

KEY WORDS

Quality   whether the product is fit for its purpose

Standards   a set of rules laid down by the client, the manufacturing sector or by law

British Standard   a set of rules laid down by the British Standards Institute (BSI)

PORTFOLIO NOTES

How will you judge the quality of your product? Are there any quality standards you will have to meet? You may need to do some research, and the British Standards Institute web site (http://bsonline.techindex.co.uk) may help you to do so.

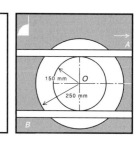

Styling Aesthetics

Have nothing in your houses that you do not know to be useful or believe to be beautiful
William Morris (1834–1896)

The aesthetics of a product are the way it looks, its style. Depending on what you are designing, the function of something may be more important than how it looks. Sometimes it can be the other way round.

If you are designing a top of the range sports car, the aesthetics of the bodywork play a vital part in how successful the car will be. Even if the engine is the best available, the car may not be seen as a luxury item simply because it does not look right.

On the other hand, the aesthetics of other parts of the car are not important. Many parts of the car will never be seen by the owner and so only need to perform as required with no consideration as to how they look.

The aesthetics may be specified by the client or they may be chosen by the person designing the product. Other, similar products on the market may also influence how a product looks.

Fig 1.10 Looks may be important on the outside...

Fig 1.11 ...but functionality is important inside

ACTIVITIES

1. Pick any two items that you use on a regular basis and list all the different things that make up its aesthetic style.

   For example, a bicycle has the following aesthetic features:

   • aerodynamics

   • colour

   • type of paint used

   • shape of handlebars.

2. Look at the two pictures below. The garments both perform the same function. State which one has the better aesthetics and try to explain why in as much detail as you can.

KEY WORD

Aesthetics   the way a product looks and feels, and the way it makes us feel when we look at it

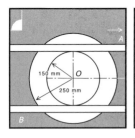

PORTFOLIO NOTES

Show your designs to other people and see how they react to the appearance of your product. Record the opinions.

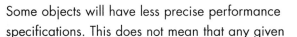

Performance

This information tells us how well the product has to perform. In some cases, this will be very specific information but, in others, it may have more flexibility. For example, the performance specification for a mountaineer's compass will be very precise. It will probably include maximum and minimum operating temperatures. It will need to be waterproof and not be affected by dirt or dust. It will have to be resistant to scratching. This information is sometimes known as the working environment.

Some objects will have less precise performance specifications. This does not mean that any given specifications can be ignored, it is just that they may not be as important in the design.

Fig 1.12 A mountaineer's compass

ACTIVITIES

1. Choose any two pieces of sporting equipment and list as many performance specifications as possible.

2. From the following list, write down which things are performance specifications and which things are not:

   - to be red

   - to work underwater

   - to look stylish

   - to work when covered in grease and oil

   - to cost about £120

   - 5000 are needed

   - to not rust.

KEY WORD

Performance how well the product has to perform under the operating conditions

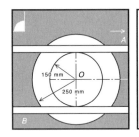

PORTFOLIO NOTES

Include a list of the performance specifications for your product.

Intended Markets

Whenever you design something you need to think about where it will be used as well as who will use it. It may be that the same product will have more than one market. A diamond tipped drill will have a very specific market but a cordless electric drill may be used by professional builders, electricians, carpenters or home D.I.Y enthusiasts. All of these will have different needs and wants from the product.

The product may be used or sold in more than one country. This may mean that any information or instructions must be available in different languages or that several sets of laws and regulations need to be considered in the final design. Sometimes it may just mean changing the name!

In most cases, engineers do not design brand new objects; most of their work is in adapting and changing existing products for various reasons (new functions, new styling, etc.). This means that they need to know about the market and what products are already available.

If you are working on a product that is part of a range from a particular company then it is important to consider how it will fit in with the rest of the range. If all the tools made by a company have yellow plastic handles then a design with a plain wooden one will not be appropriate. If similar products have a common style across a range then this is sometimes called a 'house style'.

If you make an object look too different from the other things on the market it may be a great selling factor but it may also mean that no one buys it. When research is carried out, this needs to be considered.

Fig 1.13 A space-age car may look too radical and may not appeal to the market

It is also important to consider other possible differences between markets. DVD players are region-coded depending on where they are sold. Many of them are made in the same factories and just altered before they are shipped to the different countries for sale. Ergonomic or statistical data used – average height, hand size, etc. – may only be relevant to one country. It may vary in different locations. This may mean designing different products for different territories or just making one or two minor changes for each country.

ACTIVITY

You are designing a mains-powered hairdryer that will be sold in the UK, Europe and the USA. Research and write a report about all the possible differences between these markets that you will need to consider for this product.

KEY WORDS

Market   the people you are trying to sell to

Compare   look at other products on the market

House style   the appearance of the products controlled by the company

PORTFOLIO NOTES

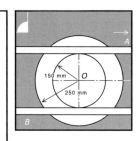

You should try to carry out market research if you can. Computer spreadsheets and databases can help you to analyse the results of research.

You should show that you have looked at other products that are already being made, and that you can analyse their good and bad points.

You should learn from other people's designs, not copy them.

Size

It is very important to specify the size of any object in three dimensions. The original design brief from the client may have rough dimensions that can be changed, but it may have specific dimensions that cannot be. In the design process, it is vital to know what you can change and what you must keep the same.

For example, it may be possible to alter the height of a wardrobe during the design process but if any nuts,

bolts or screws are used then the guide holes will need to be a fixed size.

You may also be given other information about the physical properties of the object, such as shape or form. These will be dealt with in more detail later.

There are many conventions about sizes and how they are shown in engineering drawings. These will be dealt with in more detail in the Engineering Drawings section starting on page 33.

ACTIVITIES

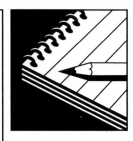

Pick any item or product and carry out the following.

1. Produce a dimension sheet for it, listing all the dimensions.

2. Make a dimensioned drawing of it.

KEY WORD

Dimension   size

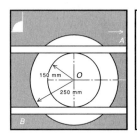
PORTFOLIO NOTES

Produce dimension sheets and drawings for your product, showing which dimensions are fixed and which you may change.

Maintenance

During their useful life some objects may never need to be cleaned or taken apart, but many things will need some kind of 'looking after' at some point.

This may just mean cleaning, or it might be that certain parts need replacing to lengthen its lifespan.

A ballpoint pen will need no maintenance during its life. A cafetière used for making coffee will need regular cleaning, so the metal filters at the bottom must unscrew from the plunger arm easily. This has nothing to do with the function of the cafetière but is required for maintenance.

Parts of a car engine will need both cleaning and replacing during its working life. If some parts wear

out more quickly than others, should you throw the whole thing away or make it possible to replace the worn parts? Sometimes high labour charges mean it is actually cheaper to replace something than repair it.

When an object needs to be dismantled for cleaning or maintenance, consideration needs to be given to how easily this can be done. The cafetière can be unscrewed without specialist tools, but a car engine may need specialist tools and personnel. This may pass on extra costs and responsibility to the customer who may not want or expect this. Small changes in design may reduce the need for specialist tools and personnel.

There are sometimes ways round this. Some companies that sell flat pack furniture include the tools required to assemble them within the pack.

Fig 1.14 A cafetière...

Fig 1.15 ...and it's many parts

ACTIVITY

Copy and complete the following table. You will need to give three further examples of products for each type of maintenance. One example has already been provided for each.

Maintenance	Product			
None	ballpoint pen			
Non-specialist	cafetière			
Specialist	car engine			

KEY WORD

Maintenance work that must be done to keep a product functioning properly

PORTFOLIO NOTES

Will your product need to be maintained? If so, you should explain what the user will have to do to keep the product working well and safely.

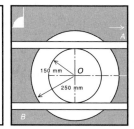

Cost

When considering how much it would take to make something, you will need to include all the different stages involved in making the product. These are:

- the design cost
- the production cost
- the cost of materials.

The design cost is likely to be fixed no matter how many items are made. It costs the same amount to design a drinks can no matter how many cans are eventually manufactured.

Fig 1.16 The design of a drinks can is a fixed cost

The production cost will usually be made up of two things. Building a machine that makes soft drink cans will cost a fixed amount but it will also cost a certain amount to run it. The running costs will include things like electricity and machine maintenance.

The cost of materials is likely to be fixed for each item made. If each can uses 3p worth of aluminium then that is a fixed cost, but it may be possible to reduce the cost if large quantities are made and materials can be bought in bulk.

The total cost is the total of these three added together (sometimes known as the factory gate cost). If you divide the total cost by the number of items you make, this gives the total cost per unit or unit cost. The design brief may specify the price or price range that an item is to be sold for rather than the total production cost.

In most cases the more units that you make the cheaper the unit cost becomes. The example opposite for a can-making machine shows how this works.

- The design cost: it costs £2500 to design the can.
- The production cost:
    - the machine to make cans costs £1500
    - the machine running costs are £250 per day
    - the machine can make 5000 cans per day.
- The cost of materials is £0.05 per can for aluminium and paint.

How much does it cost to make 10,000 cans?

Design cost	= £2500
Production cost = £1500 + £500	= £2000
(the machine needs to run for two days)	
Material costs = 10,000 x £0.05	= £500
Total cost = 2500 + 2000 + 500	= £5000
Unit cost = 5000/10,000	= £0.50 per can

ACTIVITIES

1. Find out the total cost and the unit cost for making:

    a. 5000 cans

    b. 50 000 cans

    c. 1,000,000 cans.

    Show your working out as in the example above.

2. How many cans do you have to make so that the cans cost:

    a. £0.20 each

    b. £0.10 each?

This is much more difficult. A computer spreadsheet can be used to experiment with costs like these by setting up a computer model. You can then experiment with questions like 'What if we make 1,000,000 cans?' and the computer will work the costs out quickly.

KEY WORDS

Total cost Set up costs + running costs + material costs

Factory gate cost the total cost of making a product

Unit cost cost of a single product

PORTFOLIO NOTES

You cannot set up in mass production, but you can think about the costs that would be involved in doing so. Would your product cost a lot to design? Does it use expensive materials?

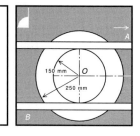

Regulations

There are two main areas where you will need to consider any relevant and appropriate regulations that may apply: manufacturing and the intended market.

Manufacturing Regulations

Wherever anyone works there are laws regarding health and safety. If you were working in an office, playing and reviewing computer games all day, there would be many laws and regulations governing your safety at work. Much of this regulation is covered in the Health and Safety at Work Act (HSWA) 1974. There may also be other, more specific, laws and regulations that will apply.

Although health and safety regulations are very important, there may be other regulations that are also relevant. Some companies and factories may have their own internal regulations. There may also be European regulations that apply.

In any production process there will be some waste. There will be issues relating to the disposal or recycling of this. A company will have to make important decisions about the environmental impact that they cause.

Market Regulations

The market where the product will be sold will mean that there may be different legislation to consider. There are different quality and safety standards for the UK, Europe and world markets. Wherever the item is to be sold, it must comply with the appropriate legislation. This may mean that a product available in one country is over-designed in order to make it acceptable in another.

Other examples of legislation are:

*   the Trade Descriptions Act: This relates to whether the description you give of a product is correct, i.e. does it do what you said it would?

*   the Sale of Goods Act: This is to make sure that all goods sold are of a saleable (or merchandisable) standard, i.e. are they fit for the purpose?

These two are separate acts of legislation but they sometimes work together. Even if a product satisfies the Trade Descriptions Act, if there are problems with it a consumer will still be protected by the Sale of Goods Act.

ACTIVITY

Put together a report describing how relevant legislation affects the design of a product.

KEY WORDS

Legislation   laws applying to a product

Health and Safety at Work Act   law to keep workers from being injured or becoming ill because of their job

Trade Descriptions Act   law that makes sure a product is described accurately

Sale of Goods Act   law that makes sure goods are fit for the purpose they were sold for

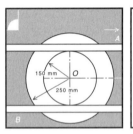

PORTFOLIO NOTES

What laws will apply to your product? Health and safety regulations will certainly apply. Are there any specific risks your product will create? Try to think about it when you design and try to keep risks to a minimum.

Production

The scale of production, the methods employed and the materials used all have an effect on each other so we shall consider them as one section.

If one wooden picture frame is needed then it is probably going to be easier to make it by hand. If 400,000 are needed, then some sort of automated production system is more likely to be suitable.

The same sawing equipment cannot be used for both wood and glass (see figure opposite). It is part of the engineer's job to consider how the product will be made. This is likely to have an influence on the material chosen.

What materials are chosen will depend on their properties and behaviour. The properties of a material are not just relevant to the final product, they also affect the process of manufacture. The materials used are often chosen as a compromise between those needed to best satisfy the required function and those most practical for production methods. It may be possible to come up with a wonderful design for a small plastic toy that fits all the criteria in the brief, but if it needs very precise and expensive machinery to make it then the design needs to be changed because the selling price of the toy will be higher than people are prepared to pay.

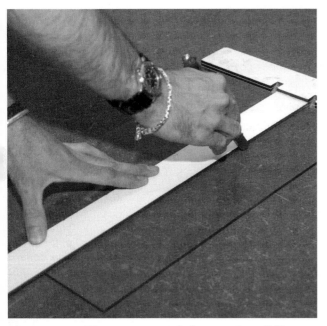 

Fig 1.17 Different materials require different cutting tools

Methods of Production

The method used to process the materials varies according to the type of product.

MASS PRODUCTION

Mass production is also known as continuous or competitive manufacturing. In mass production the product is moved continuously through the manufacturing process a bit like water flowing down a river.

Mass production is ideal when you need a large number of identical products. Most workers in mass production processes are specialized in their job. This means that they are trained to do one job quickly and efficiently. The name given to this type of worker is semi-skilled or unskilled as their skills are very narrowly-defined and specific to one task.

Workers doing these jobs often get bored and inattentive. In order to prevent this, modern mass production systems often arrange workers in teams, and then train them so that they can switch between jobs. The teamwork also increases job satisfaction.

Computers have taken over a lot of the repetitive jobs. There are two types of computer automation:

- Specialized equipment, which, like the worker, is designed to perform a single job in the production process.

- Computerized production that is programmable to do different jobs.

In modern production there is a mixture of human and machine processes and often computer-aided design (CAD) systems integrate with computer-aided manufacturing (CAM) systems to provide what is called computer-integrated manufacturing (CIM).

BATCH PRODUCTION

Batch production is also known as intermittent manufacturing, flexible manufacturing, job lot manufacturing or jobbing. In this type of manufacturing process, a batch of products is made at a time and then the equipment is set up to make a different product. Batch production is ideal where a company needs to produce a large number of different products, but it is harder to plan than mass production. Getting the right materials at the right time is crucial for efficient production to take place. Companies often use planning tools such as Material Requirements Planning (MRP) or Manufacture Resource Planning (MRP).

The advantage of batch production is that the same production line can be used to make different products without investing in different tools, people and machinery.

Custom Manufacturing

In custom (or one-off) manufacturing, each product made is unique. One person or several people may work on a product from beginning to end. Workers are usually highly skilled. Problem solving is an essential part of custom manufacturing. Products that are custom-made are usually highly priced and are often made for a specific customer, known as a guaranteed buyer.

JUST IN TIME MANUFACTURING (JIT)

JIT is not a manufacturing system in itself, it can be applied to mass, batch or custom manufacturing. The JIT approach is simple:

- produce a product just in time for it to be sold
- produce fabricated parts just in time for their use
- receive deliveries of materials and components just in time.

The advantage of this system is that it saves on storage and warehouse costs. The Japanese, who initially integrated this type of system into their production processes, called the process KANBAN. Stockless production is the name often given to JIT in Europe and America.

Another advantage of JIT is that it reduces lot sizes. A lot size is the size of the batch of the materials or components that a manufacturer receives from outside suppliers. This reduces the chance of faulty parts (known as defective parts) from being hidden in a large batch. For example, it is easier to notice five faulty parts in a batch of 50, than it is to notice them in a batch of 50,000.

JIT reduces costs in other ways too. When a manufacturer orders materials they have to pay the supplier, but the manufacturer does not receive any income until they sell the finished goods. If they can reduce the amount of money they have to spend before receiving an income then they can reduce the overall costs of production.

JIT is a careful balance between purchasing and production because a manufacturer does not want workers waiting for components to arrive. How much to purchase is known as the economic order quantity (EOQ).

FLEXIBLE MANUFACTURING

Flexible manufacturing lets a company make a wide range of variations of the same product to meet specific customer requirements. For example, a motor manufacturer can produce cars with different engine sizes, tyres, colours, etc., all to specific customer requirements. Flexible manufacturing relies upon a large amount of automation and computer control.

SUMMARY

If I were making 200 hot cross buns to sell at a school fête, then this would be a batch process. I could then use the equipment to make Christmas cakes.

If I were setting up as a baker planning to make hot cross buns for the rest of my life, then this would be a continuous process.

Fig 1.18 How many hot cross buns will you need?

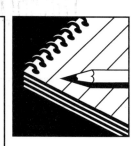

ACTIVITY

For each of the following situations say whether you think it would be a batch or continuous process:

1. making sandwiches for a family party

2. producing replica shirts for a premier division football club

3. making Formula One racing cars.

KEY WORDS

Mass production   making large numbers of identical products

Custom manufacture   making a single product for a specific customer

Just in Time manufacturing   ordering materials so that they arrive when they are ready to be used and making products when they are ready to be delivered

Flexible manufacturing   making different variations of the same product

Batch production   making a batch of the same product before moving to the next product

PORTFOLIO NOTES

If your product went into full production, what type of production methods do you think would be suitable?

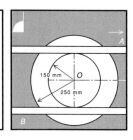

Scale of Production

The numbers of a particular product required could have a very significant effect on the process by which they are manufactured. If a relatively small and fixed number of something is required, then it may be financially more sensible to adapt a process or a piece of machinery that already exists.

If a much larger number is needed then it will be more sensible to have a dedicated production system. Some of the things that could be considered are listed below.

- Is extra machinery needed?

- Can existing machinery be adapted?

- Are there any knock-on effects on other products?

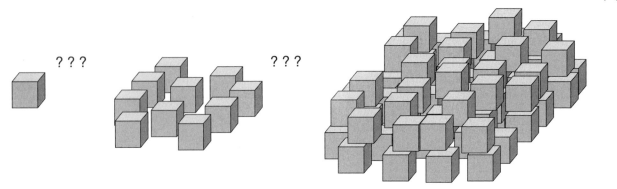

Fig 1.19 The quantity required will determine the production method used

- Can existing parts be bought in?
- Is more staff needed?
- Will there be further orders?

The cost of all possible options will usually be the deciding factor in these types of decisions.

③ DESIGN SPECIFICATIONS AND SOLUTIONS

Research and Analysis of Information and Data

The purpose of research is to get a fuller understanding of all the aspects of a problem. The information that you find will sometimes help you with possible solutions.

Questionnaires and surveys are commonly used to give an understanding of the market for a particular product. Some of the things you might find out include:

- what is the market at present?
- what is the potential market?

- what other products available provide a similar function?
- what are the requirements, expectations and desires of possible users of the product?

This is called market research. The data collected needs to be analysed so that it can provide information for the design process.

When a client provides you with the design brief, they may have already carried out some initial research. If this is the case then you may just have to analyse the data they give you. Otherwise you will need to collect information that will help your design work.

Fig 1.20 Take a survey to gather information

What to Ask?

There are several different ways that you can collect information but this usually involves asking specific questions of a group of people.

Questions can be either open or closed.

- Closed questions have a simple fixed answer (e.g. yes/no, bigger/smaller).

- Open questions can be answered in many different ways (e.g. how do you feel about...?).

Closed answers are much easier to analyse, but open answers will give you more information. You will need to choose what you ask carefully in order to get information that is useful. Closed questions do not always just have two answers, for example:

What colour car do you drive?
RED   BLUE   WHITE   BLACK   GREEN   OTHER

Analysing the data can be very time consuming so make sure you only ask questions that will really help you. Using tick boxes for closed questions will make it quicker and easier. Computer software such as spreadsheets and databases can help you analyse your market research.

Who to Ask?

The group of people you can ask could be random or targeted:

- random – no selection involved, you could literally send a questionnaire to 1000 people chosen at random from the electoral roll.

- targeted – you decide on a specific group and only ask them; groups could be boys aged between 8 and 12, people who drink tea, people who are members of a political party, etc.

Again you need to think about what information you want. If you are designing a car there is not much sense in asking people under 16.

Whatever questions you choose to ask and whoever you ask them of you must make sure that your data is valid. This includes always asking the same questions and not hinting at or suggesting a particular answer. It will also mean that you must ask enough people to get a fair sample.

Sometimes it is necessary to give people a reason to reply otherwise many may not bother. This could be done by entering all names into a draw for a prize such as giving out discount vouchers for local shops.

It is also important to ask a representative selection of the sort of people who might buy your product. If you only ask people in a wealthy area how much they would pay for a product, the answers they give will not be the same as people from a poorer area.

Please fill out this questionnaire

Send your answers in and get a £2.50 voucher for Bob's Burgers

Fig 1.21 Special offers may encourage feedback!

ACTIVITIES

1. For each of the following say whether they are open or closed questions.

    a. Do you like cheese?

    b. Would you like a square or a round table?

    c. Are you happy?

    d. What makes you happy?

    e. What do you want the product to look like?

    f. What three things do you like the best about the sample?

2. Write down at least five questions that you could ask which would help you if you were designing a new bicycle. You must include open and closed questions and say which are which.

Other Research

Not all the information that you will need can be obtained by asking people questions. This will help you understand the market for the product but it will not be enough in itself. Other things that you may need to research could include:

- relevant legislation and regulations

- available manufacturing processes

- available materials for manufacture

- the properties of those materials

- similar products available.

Many of these are covered in more detail elsewhere in this book.

KEY WORDS

Market research   finding out about the potential market for a product

Analysis   looking for patterns in the responses that will help you decide what to produce

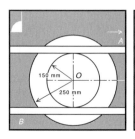

PORTFOLIO NOTES

You should carry out some market research for your ideas before you decide exactly what to design. Results can be calculated by computer and charts may help you see patterns in the data.

How to Present Your Data

It is important to display the data you have collected in order to support any decisions that you may make based on it. Most of the data collected from questionnaires or surveys can be shown in pie or bar charts. These types of graphs and charts are good, clear ways of showing information but only work well for closed questions.

If you use these charts it is important to make sure that they are clearly labelled and the important information is easy to see. Figure 1.22 below shows an example of the same data shown in two different ways. Although the data is the same, the second table is much easier to read.

CARS	
VW	46
VOLVO	18%
23% SKODA	
BMW	3
AUDI	°/100

WHAT CAR DO YOU DRIVE?	
MAKE	PERCENTAGE
VW	46
VOLVO	18
SKODA	23
BMW	3
AUDI	10

Fig 1.22 Data is easier to interpret when presented clearly

Scientific Principles

Science gives us rules and explanations about how things behave. Scientific understanding provides a foundation for all engineering. It is impossible to design anything without considering some scientific ideas.

In any design work you must first identify what the relevant laws and rules are, and then make appropriate calculations. You will need to include this information in your design. By applying the correct scientific ideas you will be able to answer these types of questions:

- how strong does it need to be?
- how hot will it get?
- will it melt?
- what is the maximum load it can take?
- will it work underwater?
- will it break or bend if I overload it?
- how fast can it go?
- what current will the motor need?

Some of these scientific ideas are easy to understand and find out about, but some are more complex. You may also find that there are data books that will help you.

The most common laws of science that will be relevant are the ones relating to forces, energy and how materials and structures behave. In order to understand how these ideas will work, there are two examples below.

Problem: You are designing a nail gun. The nail must accelerate out of the gun at 150 m/s² in order for it to work. What force must the gun put on the nail for this to happen? The nail has a mass of 0.05kg.

Solution: The scientific idea here is the relationship between force, mass and acceleration (one of Newton's laws). The equation is:

Force (N) = Mass (kg) x Acceleration (m/s²)

So, putting the numbers in:

$$F = M \times A$$
$$F = 0.05 \times 150$$
F = 7.5 Newtons

When you apply scientific ideas it is not always just a case of putting numbers into equations, sometimes it is about understanding other ideas.

Problem: If there is a minor earthquake, some tall buildings are at risk of falling down. How can they be made more stable without rebuilding them?

Solution: To solve this problem you need to look at the scientific ideas relating to structures and how they can be strengthened. You also need to know about earthquake waves. This can be a very complicated area but the following key ideas are relevant to this problem.

- Some materials and shapes are strong in one direction but not in another.

- Concrete is very strong in compression but weak in shear.

- Earthquake waves move objects up and down and from side to side.

- The concrete in the building should withstand the up and down earthquake waves because it is strong in compression, but is likely to be damaged by the side to side waves as it is weak in shear.

The outside of the building can be fixed with X-shaped supports. This means that when the building is moved from side to side the arms will take the force and not the concrete. This should protect the building.

This method has been used on some buildings in Los Angeles and San Francisco that are at risk of earthquakes.

Generation of Ideas and Solutions

Once you have identified the problems, collected all the information that you can get and done all the necessary research, then you have to start coming up with some ideas.

It is very likely that as you have reached this stage you will already have had quite a few ideas, but it is important to try and get as many as possible before you start work in more detail on one of them.

Fig 1.24 Mind showering

Fig 1.23 Developing an idea

If you are working on your own it is very important to try and have a system so that all ideas are recorded. In some cases this may just be a scribble on the back of an envelope but, as a professional, you should try and formalize it a bit more.

Try to include a number of people in the initial idea phase. If you do it on your own it is possible to have what you think is the 'best idea' stuck in your head and this stops you from thinking about alternatives.

There are many ways of generating ideas, some are more formalized than others but they all work in a very similar way. Some of the common names are brainstorms, mind maps, spider diagrams, idea webs and mind showering.

Mind showering involves writing down any idea that comes into anyone's head without really worrying about how useful they may be. As the name implies, mind showering is just a way of getting every possible idea down on paper (see Figure 1.24).

Mind maps, spider diagrams and idea webs are really just different names for the same thing. They are different from mind showering because they are more formal and have to show a connection between thoughts (which is why they are called maps, diagrams or webs). This makes it easier for someone who looks at them later to understand.

To begin with you would write the problem down on a piece of paper, then write down all the different parts of the problem that need answers. You may well find you need to keep adding bits as you go along.

It is much easier to see this with an example. Figure 1.25 (overleaf) shows a simplified mind map for a bedside lamp. Some solutions have been decided upon and others have still to be agreed.

Evaluation of Ideas

It may be that when you come up with ideas you think that you have got 'the one' that matches the design brief perfectly. This does not mean that you can go ahead and make it, you need to evaluate all ideas in a fair way before you decide which one will move on to the next stage.

The easiest way to try and evaluate all the ideas that you have come up with is by using something called an evaluation matrix. This is a table where you list all the design criteria down the side of a table and all the possible solutions along the top. In each box you then put a mark or score describing how well each solution matches the criteria. Below is an example of an evaluation matrix.

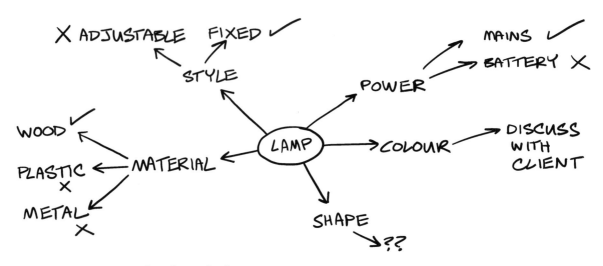

Figure 1.25 An example of a mind map

Design Criteria	Possible Solutions			
	1	**2**	**3**	**4**
A	+	+	−	−
B	+	−	+	+
C	+	+	+	−
D	−	−	−	+
E	+	**N**	**N**	−

Key:
+ matches criteria
− does not match criteria
N not relevant

Look at the table and use this to help you decide. Here the one with the most pluses would be chosen (solution 1) but if you used some type of scoring system (e.g. 1 = bad, 10 = good) then you would total up each possible solution and pick the one with the highest total. It is up to you to decide how you want to set up your matrix.

Once you have what you think is the best idea you will need to test and evaluate it further. This may mean making a prototype and testing it, or testing individual components. Whatever happens next, the results of any testing must be fed back into the system so that any modifications or changes can be made. It is very rare for a first possible solution to make it through to production without any modifications.

ACTIVITY

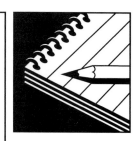

Copy and fill in the evaluation matrix below for these personal stereo solutions to find the best one.

	Mass (kg)	Power	Material of construction	Cost (£)
Solution 1	10	mains	cast iron	10
Solution 2	0.5	battery	plastic	25
Solution 3	0.2	battery	polystyrene	10

Design Criteria	Possible Solutions		
	1	**2**	**3**
Portability			
Lightweight			
Hard Wearing			
Affordability			

KEY WORDS

Brainstorms, mind maps, spider diagrams, idea webs and mind showering   ways of recording your initial ideas

PORTFOLIO NOTES

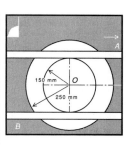

You should include copies of your mind maps and any other records of your initial ideas. Don't worry if they look untidy, they are supposed to be produced quickly to help your thought process.

You can then draw up an evaluation matrix to explain why you chose the idea you think is best.

Modelling Techniques

Once you have decided which solution is the best one to put forward, you will then need to make sure that it does actually satisfy the design criteria. There are several ways of doing this:

- using prototypes – clay or lightweight material

- using scale or full-size models

- using ICT modelling and CAD.

If you want to get an idea of how something will look and behave you can build a prototype. This will be based on a preliminary design and may not have all the features of the final product. These models are usually made from lightweight materials like polystyrene.

Scale and full-size models are more sophisticated. They are made to a more precise set of details and can be used in testing for specific behaviour (e.g. strength or aerodynamic properties). They are often fully functional.

As computer processing power continues to increase it is possible to perform more and more sophisticated modelling. You enter all your available data into a computer system and then you can check and test how it behaves. This is dealt with in much more detail in Chapter 3.

Fig 1.26 A London bus (left) and a scale model of a London bus (right)

ACTIVITIES

1. For each of the following situations, state which would be a suitable modelling technique and why.

   a. Which shape of mobile phone is most comfortable to hold?

   b. How does a rocket behave in space?

   c. What car shape is the most aerodynamic?

2. Research each of the modelling techniques listed in this section, trying to find out where they are used.

KEY WORD

Modelling exploring an idea in three dimensions, either in modelling materials or by computer

PORTFOLIO NOTES

What modelling techniques would help you visualize your product? Take photos of any real models you build and include them in your portfolio. If you use computer models, print them and include them.

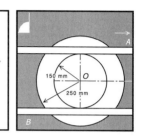

ENGINEERING DRAWINGS

A sheet of paper has only two dimensions, length and width. Its thickness is insignificant. The solid objects you design will have length, width and depth. You will therefore need a way of representing these.

Artists have their own way of representing solid objects on paper. They use drawings or freehand sketches. Engineers and manufacturers also use freehand sketching techniques, but because detail and accuracy are very important they have a range of specialist representation techniques, the most important of which is orthographic projection.

Many engineering drawing work involves trying to draw three-dimensional (3D) objects on a two-dimensional (2D) piece of paper. This can be done in several ways. There are many 2D and 3D drawing and sketching techniques. You will have to decide which technique to use when. Some of the techniques you will need to know about include:

- freehand sketching
- perspective drawing
- first and third angle orthographic projection
- isometric projection

- oblique projection
- assembly diagrams
- exploded diagrams.

They are all covered in more detail on the following pages. You will also be expected to be familiar with CAD software and hardware systems. This is dealt with elsewhere.

Freehand Sketching

By learning a few basic drawing techniques you can make it possible to draw both good and useful freehand sketches. You may have an idea and want to get it down on paper very quickly and so you need to be able to make it a bit more than just a scribble or a doodle.

Freehand sketches will never replace engineering drawings. This does not matter because they do different jobs. Freehand sketches are just a way of getting initial ideas on paper.

All simple shapes are built up from either straight or curved lines. Although it may sound a little silly you should practise drawing both straight and curved lines. A little practice will improve your technique considerably.

- Try drawing a straight line from one side of an A4 piece of paper to another. It helps not to look at the point of the pencil, but at where it is going.

- Practise drawing circles by drawing a square first. A circle will just touch the edges at the middle of each side. Drawing lines at these points will help.

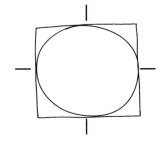

Fig 1.27 A freehand sketch of a circle

ACTIVITY

Pick any product available to you and draw a freehand sketch of it.

Make sure you have access to this product, as you will need it for many of the drawing questions that come up in the following sections.

Perspective Drawing

This will give you a good idea of what the object will look like, but is not very useful for giving specific information that can be used in design and manufacture.

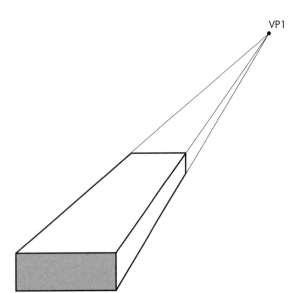

Fig 1.28 One point perspective

In perspective drawing, all lines moving away from the viewpoint converge or meet at vanishing points.

These can be in the foreground or background. There can be more than one vanishing point but they all must be on the same line called the horizon.

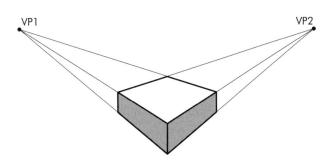

Fig 1.29 Two point perspective

This is because the further away objects are from you the smaller they appear to get. If you were to stand in the desert and ask somebody to walk away from you, they would seem to become smaller and smaller until eventually they vanished. The name given to the point at which they vanish is called the vanishing point. Imagine yourself at one end of the classroom, in front of a glass panel. If you trace what you can see onto the glass panel, you will construct a perspective drawing. The tables nearer to you look bigger than the tables that are further away. Artists call the glass panel 'the picture plane'.

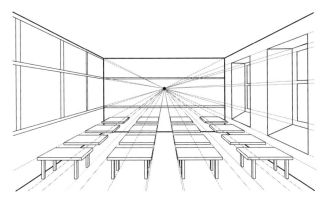

Fig 1.30 A classroom with one vanishing point

In the example given above, all of the tables are in nice regular patterns. In reality, they would be at different angles and there would therefore be more than one vanishing point.

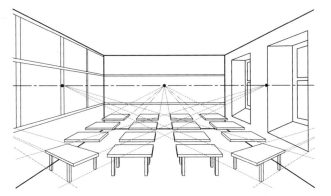

Fig 1.31 A classroom with more than one vanishing point

If you move, the vanishing point moves with you. The other problem with our example is that all of the objects are flat whereas, in real life, objects slope uphill and downhill relative to the horizon.

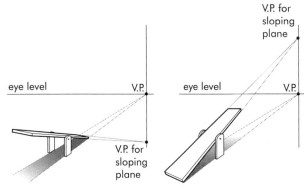

Figure 1.32 Vanishing points of a sloping plane

If you need to produce accurate freehand views, you have to add a third vanishing point. Imagine viewing a tower block from below or above. The end that is furthest away is smaller than the end that is closest.

ACTIVITY

Draw again the object you drew in the freehand sketching section, but this time as a:

1. perspective drawing with one vanishing point

2. perspective drawing with two vanishing points.

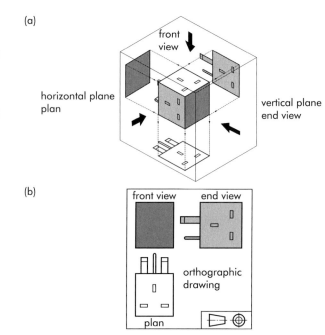

Orthographic Projection

Orthographic projection is a formal technical drawing technique usually having three views, drawn to scale and at 90° to each other. Orthographic projection is usually shown by drawing three views: a front view, a plan and an end view (called front, plan and end elevations). Sometimes a second end elevation is added and, where even more detail is required, the three-dimensional object is shown as if it were sliced or cut.

Orthographic projection is usually shown in what is known as 'third angle projection'. The difference between third angle and first angle projection is shown in the following illustrations:

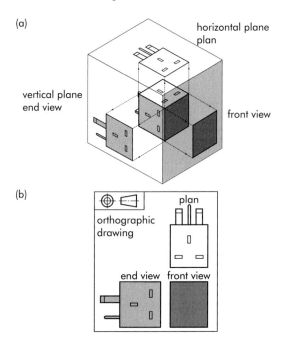

Figure 1.33 Third angle projection

Fig 1.34 First angle protection

As can be seen there are two ways of drawing an orthographic projection:

• first angle or English

• third angle or American.

The only difference between the two is the way the final drawing is set out. If you are able to draw one way then you can draw both.

Orthographic drawing layouts are covered by the British Standards BS308, PD7308 and PP7308, which give guidelines and symbols.

Orthographic drawings may not give you a very clear idea as to what the finished object will look like, but they need to be precise for the production mechanism.

Third Angle Orthographic Projection

This is the most commonly used orthographic projection. The layout of the drawing is:

- plan elevation

- side elevation under plan elevation

- end elevation to left of side elevation.

The easiest way to understand how one works is to look at this example:

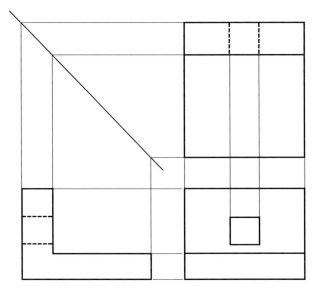

Fig 1.35 Third angle projection

How to draw a third angle orthographic projection:

- Plan: look directly down on the top of the component and draw what you see.

- Side elevation: look directly at the side of the component, drawing what you see.

- End elevation: look directly at the end of the component, drawing what you see.

You can see in the example above that construction lines can be drawn in to help.

First Angle Orthographic Projection

The way to draw a first angle projection is the same but the layout is:

- side elevation

- plan elevation under side elevation

- front elevation to right of side elevation.

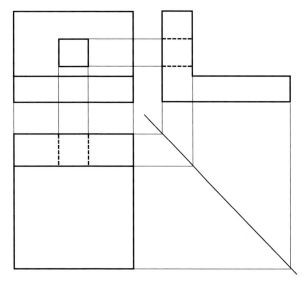

Fig 1.36 First angle projection

Useful Rules and Conventions For Orthographic Drawings

- Dimensions are always in millimetres.

- Dimensions are written above the dimension line.

- Vertical dimensions need to be read when the drawing is turned on its side, so put them on the left-hand side of the vertical dimension lines.

The projection lines should not touch the object.

ACTIVITY

Using the object you drew in the freehand and perspective sections draw it again but this time as:

1. a third angle orthographic projection

2. a first angle orthographic projection.

Pictorial Views

There are two pictorial views that combine the visuals of sketching and perspective drawing, and the information contained in orthographic projections. In both cases, true measurements can be taken from the diagrams, which are called isometric and oblique projections. They may look similar but there are some important differences.

Isometric Projection

Isometric drawing was invented by engineers as a way of representing three dimensions using a traditional drawing board. Whilst isometric drawing does not give an accurate visual representation of a product, it is commonly used by engineers and manufacturers.

An isometric diagram is drawn with the aid of two isometric axes. These are simply two lines that start at a point at the front of the object and rise up at 30° to the horizontal. Sometimes isometric graph paper can be used.

Rules for isometric drawing:

• all horizontal lines are at 30° to the horizontal

• all vertical lines remain vertical

• all measurements are either true or to scale.

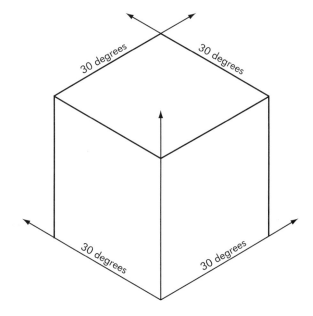

Fig 1.37 Isometric projection

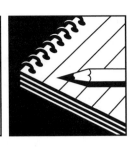

ACTIVITY

Using the object you drew in the previous activities, draw it again but this time as an isometric projection.

Oblique Projection

Another drawing technique is called oblique drawing. Oblique drawing is another method used by engineers to represent three-dimensional objects on paper.

You begin by drawing a true view, to scale, of the most difficult or complicated side. You then take a line at 45° to the horizontal (called a receder) and use this to form the other views.

Taking a regular block as an example, one face of the block is drawn to the correct size and the third dimension is shown as a sloping line usually drawn at 45°. There is no unique angle or scale.

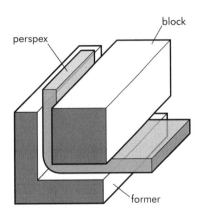

Fig 1.38 Oblique projection

All measurements on the first face drawn are to scale. The other measurements are at half-scale (shorter). This gives as realistic a view as the perspective view does, but can also give accurate dimensions.

Remember that some of the dimensions are half-scale otherwise you may end up with a half-size final product.

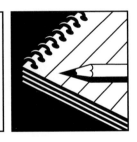

ACTIVITY

Using the object you drew in the previous activities, draw it again but this time as an oblique projection.

Assembly and Exploded Diagrams

These two are very similar types of diagram and many people consider them to be the same thing. Their job is to show all the component parts of an object and how they fit together.

An assembly diagram will be one or more diagrams (a sequence) showing how to put something together. The most common example is flat pack furniture. Inside the packet you will get a single diagram or a series of diagrams showing you how to assemble a piece of furniture from the component parts, including the fixings needed.

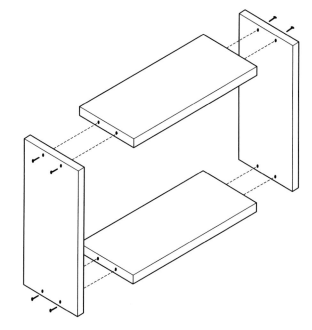

Fig 1.39 An assembly diagram

Exploded views are used whenever the designer wants to show how various parts of an object fit together. The object is drawn in isometrical perspective as if it has been blown apart.

Fig 1.40 Exploded view of a pen

The exploded view is usually a single diagram showing all the components of an object and how they fit together. It is generally used for maintenance purposes.

Rear hub breakdown

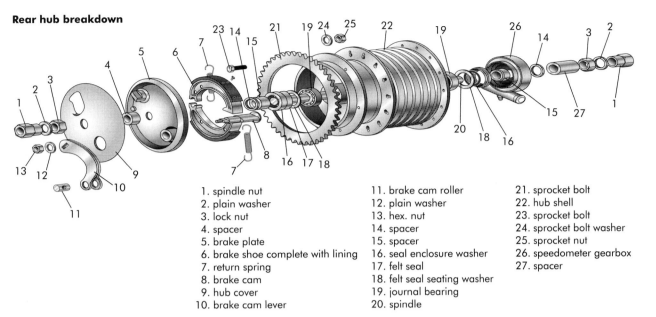

1. spindle nut	11. brake cam roller	21. sprocket bolt
2. plain washer	12. plain washer	22. hub shell
3. lock nut	13. hex. nut	23. sprocket bolt
4. spacer	14. spacer	24. sprocket bolt washer
5. brake plate	15. spacer	25. sprocket nut
6. brake shoe complete with lining	16. seal enclosure washer	26. speedometer gearbox
7. return spring	17. felt seal	27. spacer
8. brake cam	18. felt seal seating washer	
9. hub cover	19. journal bearing	
10. brake cam lever	20. spindle	

Fig 1.41 Exploded view of a rear hub with parts list

ACTIVITY

Choose one or more of the products below and draw an exploded view of it:

- a ballpoint pen
- a pencil sharpener
- a CD case.

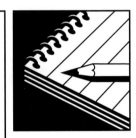

Making Your Drawings Look Life-like

Shading is used to make three-dimensional drawings look lifelike. In order to make your objects look solid, you will first need to work out the direction of the light. There are various ways of adding shading to your drawings. These are shown overleaf.

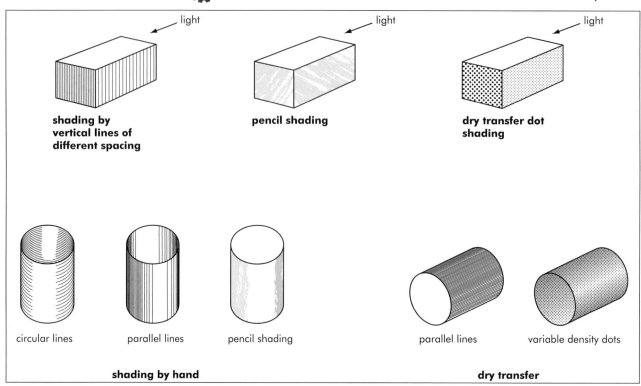

Fig 1.42 Different types of shading

Charts

Another form of drawing that you may need to do is to produce charts. Charts are used to represent figures and statistics graphically. There are three basic types of chart: bar chart, pie chart, fever chart or table.

Bar Charts

A bar chart represents each quantity as a bar or column, which corresponds to the height or length that represents the quantity to be shown. It is called a bar chart because, in its simplest form, it uses bars to represent figures. Figure 1.43 is in two dimensions, but lots of three-dimensional bar charts, like Figure 1.44 opposite, can be found in commercial use.

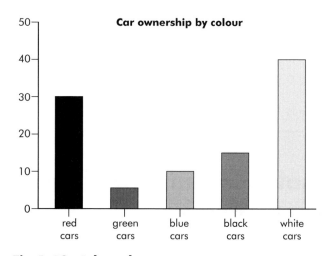

Fig 1.43 A bar chart

Bar charts are good for showing individual figures and for comparing different values. They can also show two or more sets of figures over the same time period. However, if there are too many numbers the bars become too thin, and bar charts are not good at showing flow over time.

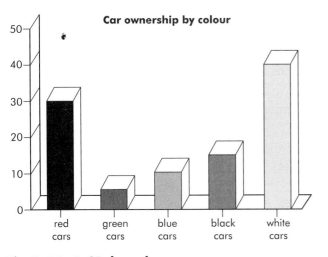

Fig 1.44 A 3D bar chart

Pie Charts

Pie charts divide the quantities into components, usually percentages, and represent these as slices of a pie. This is where it gets its name.

Car ownership by colour

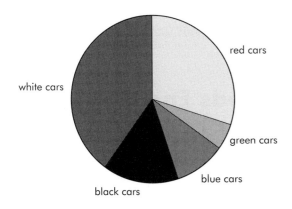

Fig 1.45 A pie chart

Pie charts are also available in two and three-dimensional forms. Pie charts are very good providing there is a maximum of eight to ten component parts. They are often used to show income, spending, budgets and time allocation. If there are too many divisions, the slices of the pie become very small making it difficult to understand.

Line Charts

Line charts (or fever charts) are similar to bar charts but have a continuous line plotted over time. They are called fever charts because they are similar to the charts used in hospitals to show a patient's temperature, rising and falling over time. The line that is plotted is usually known as the curve.

Fever lines are used to show things like unemployment figures, stock market share prices over time, temperature changes, etc. They are not a good way of showing information in which there is very little change over time.

One of the main uses for fever charts is to plot fixed and variable costs.

FIXED COSTS

These costs stay the same, are fixed, no matter how many products a company makes or sells. These costs include money spent on permanent investments such as property, machines, tools and furniture. Fixed costs also include people who work in the company but who do not make a product, e.g. the accountant, lawyer, etc. The relationship between fixed costs and the number of products made is very important as the number of products made affects the profit a company makes.

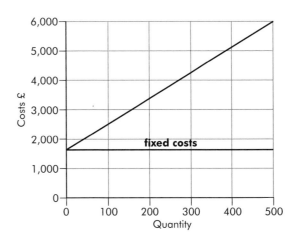

Fig 1.46 Cost graph (fixed and variable)

VARIABLE COSTS

Variable costs go up and down with the amount of products made. This is because variable costs include all the expenses associated with making the product. These will include wages for the engineers, the price of materials and the money spent on tools and supplies to make the product. If a company decides to make more products, the amount of materials used will go up and therefore the variable costs increase. If a company decides to make fewer products, the variable costs decrease.

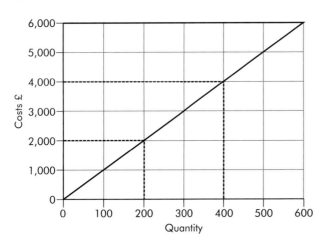

Fig 1.47 Cost and quantity graph

It is important to note the relationship between the number of products made and the variable and fixed costs. For example, if the number of products made is doubled or tripled, the total variable cost will also double or triple, but the fixed cost will remain exactly the same.

Tables

Tables simply display numbers, words or pictures in columns and rows.

	red cars	green cars	blue cars	black cars	white cars
1999	40	26	8	14	35
2000	35	52	44	75	16
2001	17	33	41	26	22
2002	32	82	92	15	47
2003	38	60	26	45	56

Fig 1.48 A table displays data clearly

Tables can be used to show family trees, timetables, calendars and flow charts. A variation of a table is a flow chart, which uses standard symbols to help understanding.

Methods Engineering Process Charts

Methods engineering involves planning the sequence of processes needed to make parts or finished products. Planners create charts to show the production process graphically.

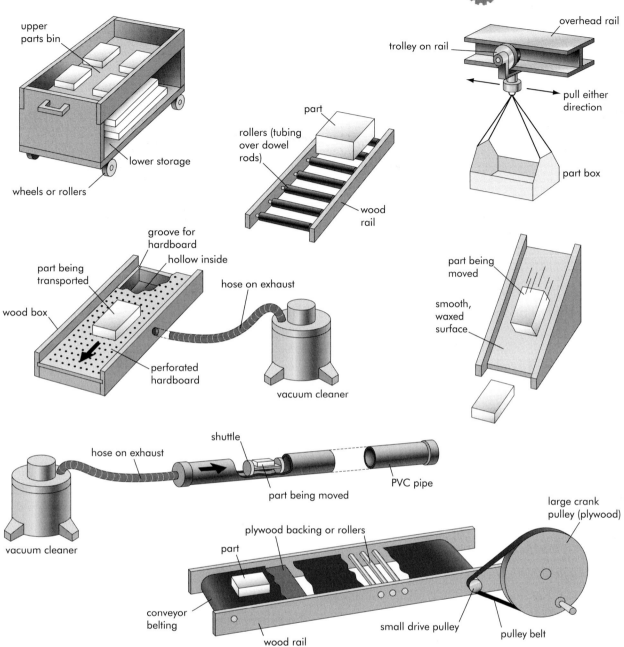

Fig 1.49 Material transport systems

A flow process chart is used to show how materials are transported and when materials or components are being stored.

In order to get a flow through the factory, manufacturers often produce other types of chart, such as plant layout charts. All of the machinery is drawn

in its location and a manufacturer will draw arrows to show the movement of the product around the factory.

There are many ways that products, materials and components can be moved around the factory (see Fig 1.49). Automated robots and conveyors are the main ways.

Other Types of Drawing

Ideogram is the name given to a range of different types of drawing used by engineers and graphic artists. The main types of ideogram used are:

- a pictogram – an image or symbol used to convey information. Road signs or hazard signs are examples of pictograms.

- a logogram (known as a logo) – a sign relating to the name of a company. Logos often consist of words and symbols.

- a monogram – a sign made up of two or more letters. Monograms are often the starting point in designing logos.

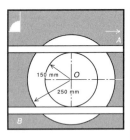

PORTFOLIO NOTES

ICT can help you to produce many of the drawings described in this section.

Spreadsheets produce tables and charts that will look good in your folder and help you to analyse figures such as market research surveys.

Object-based drawing packages can be used to draw floor plans, and may have special symbol libraries that have the shapes you need ready-prepared.

These days, even word processing and desktop publishing packages are capable of producing most of the diagrams you need.

Remember that pictures are often much easier to follow than lots of words, but they do need to be well presented.

Dimensions

In the design process, a great deal of time and effort may be spent making sure that something is the right size for its intended use. If this information is not communicated clearly and correctly in the appropriate diagrams, then all this time is wasted.

A scale diagram, correctly drawn, can show most of the required information but there will be some things that need to be added to any diagram to make sure the fullest information is given. These include:

- tolerances
- radii (internal and external)
- centres
- springs
- holes
- screw threads (internal and external).

ACTIVITY

Using the object you drew in the previous activities, draw it again but this time as a dimensional drawing. Try to include all the details you can.

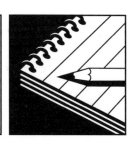

PORTFOLIO NOTES

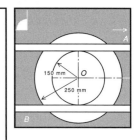

Think carefully about what drawing types you use in your work. You are using them for two main reasons: to show what the finished product looks like and to show how to make it.

Manufacturing drawings need to stick closely to the rules; presentation drawings can be much more imaginative.

Try to use a variety of drawing styles to show off your range of skills. Include technical drawings that show how your product will be made. These could be done by hand or on a CAD package on computer.

Drawings should be fully-dimensioned. Choose a sensible scale for bigger objects. Include a parts list if it is appropriate to do so.

Sometimes a rendered perspective view can show off your product really well. If your product is a complicated assembly, an exploded view might make it clearer.

Tolerances

It is very hard to make anything to exactly the size specified and, even if it were possible, it may be very expensive to do so. A toleranced dimension states what the acceptable range of that dimension is. This will be shown as either a maximum and minimum value or a plus (+) and minus (–) value. In each of the cases below the range of possible sizes is the same.

- 200 mm +/– 8 mm

- max = 208 mm min = 192 mm

- 192 mm + 16 mm/– 0 mm

- 198 mm +10 mm/– 6 mm

The tolerances specified in a design can often have an influence on the manufacturing process and materials chosen.

Although tolerances may be specified in the design there may be legislation covering required tolerances for some objects, e.g. bridges, car engines, train tracks.

ACTIVITIES

1. For each of the following, state the largest and smallest allowed dimensions:

   - 2000 mm +/− 15 mm

   - 12 mm + 2 mm

   - 250 mm + 25 mm/− 20 mm

   - 14.5 mm +/− 0

   - 200 mm +/− 10%.

2. A fence panel can be between 2500 mm and 2300 mm. Write this down as a tolerance measurement in three different ways.

KEY WORD

Tolerance   the amount of error that is allowed when a part is being made

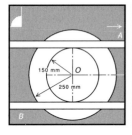

PORTFOLIO NOTES

Consider what tolerances the parts you use will need. Make them too small and the part becomes hard to make and costs more. Make them too generous and the parts may not fit or do their job well.

Radii

One of the most common mistakes people make when dealing with circles is mixing up diameter and radius. Whenever you are putting down information about any circle you must make sure it is clear and unambiguous. When a radius is shown on a diagram the number is prefixed by a letter r. When a diameter is shown on a diagram it is prefixed by a d or the symbol Ø.

Depending upon the diagram both internal and external sizes may need to be shown. The following diagrams show most examples that you will encounter.

In the case of any circular or part-circular shape, the convention is to show the centre of the circle with a crossed pair of lines. This creates a point from which measurements can be taken if needed. These lines should be broken.

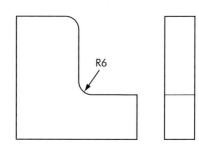

Fig 1.50 Internal radius

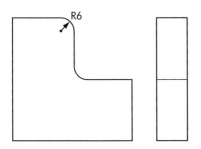

Fig 1.51 External radius

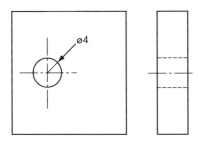

Fig 1.52 Internal diameter

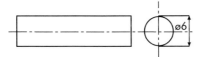

Fig 1.53 External diameter

Screw Threads

Many products will be constructed using threaded fixings, such as nuts and bolts. To show these on a drawing, a certain set of rules must be followed, so that the components are not confused with other items.

The side elevation rules are as follows.

- An external screw thread, such as a bolt, is shown by drawing a rectangular side view of the bolt.

- The threaded part is identified by drawing a line inside the bolt outline, to signify the depth of the thread. As bolts are cylindrical, they must also have a centre line drawn along their length.

- An internal screw thread, such as a nut or threaded hole, is drawn as a rectangle, with two internal lines to signify the depth of the thread, and a centreline.

The difference is that the external thread will be a separate component, whereas the internal thread will be enclosed within a block – the nut.

End elevation rules are basically the same, except that the shape of the screw thread is a circle; external threads have a broken circle drawn inside the diameter of the bolt to signify the thread depth; and internal threads have a broken circle drawn outside the hole to signify that the hole is slightly smaller than the thread (otherwise the bolt would fall through the hole).

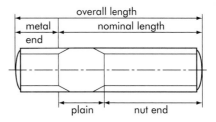

external screw threads

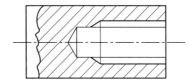

internal screw threads

Fig 1.54 Screw threads

Note: Any component that is to be threaded will not have hatched lines shown on a section drawing.

ACTIVITY

Collect a nut and bolt pair. Produce an accurate engineering drawing of the pair.

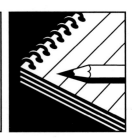

KEY WORDS

Radius distance from the centre of a circle to the perimeter

Radii more than one radius

Diameter twice the radius

Thread the ridged spiral pattern on a screwed part

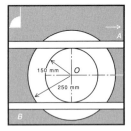

PORTFOLIO NOTES

Threads and circles need to be dimensioned accurately in your drawings, and they need to follow standard drawing conventions as described here.

Springs

Springs can be found in many engineered products. They are often used to allow movement within a device, such as a valve. Generally a spring is used to force apart two components and a second fixing, such as a nut and bolt, is used to force the same two components together. This arrangement means that the device is kept in equilibrium, until something happens to tip the balance in favour of either the spring or the nut and bolt. Such an arrangement can be found in a release valve.

As pressure increases within the system, the spring is compressed and eventually the valve is opened allowing the pressure to escape. If the escape occurs at too low a pressure, the bolt can be tightened and the spring is therefore harder to compress. If the pressure escape happens at too high a pressure the bolt can be released and the spring is therefore easier to compress.

As a spring is basically a coil, or helix, the method of drawing one is to draw a helix, although in many applications the number of coils on the spring is not known so a careful sketch would be used to signify that a spring is to be inserted.

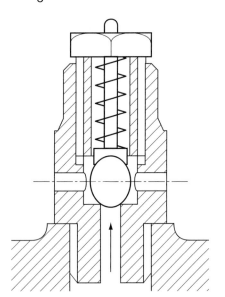

Fig 1.55 Sectional view showing spring

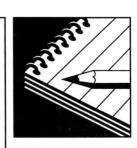

ACTIVITY

Many ballpoint pens have small springs inside the shaft to force the ink reservoir and nib to retract when not needed.

Collect a pen that you can take apart. Carefully remove the components and produce a sketch of how they all fit together – an assembly drawing – clearly showing the position of the spring.

Holes

Engineered products are full of holes!

> The wheel's hub holds thirty spokes
> Utility depends upon the hole through the hub,
> A potter's clay forms a vessel.
> It is the space within that serves.
> A house is built with solid walls
> The nothingness of window and door alone,
> renders it usable,
> That which exists may be transformed
> What is non-existent has boundless uses.
>
> *Lao-Tse*

As much of what is made is made from a number of components, there will invariably be lots of holes to house them or to locate the fixings. Generally they will be circular; as most mechanical cutting devices spin, it is a lot easier to cut a circular hole.

When producing engineering drawings it is important to signify what is a hole, and what is not. It is also essential to describe clearly the size of the hole and whether it passes right through the material or only part way through.

From the outside of a piece of material most holes will be shown as circles. They must be drawn accurately, in terms of both size and location. Imagine the difficulty an engineer would have in making something if the holes for the bolts were different sizes or in the wrong places – the person that had produced the drawing would not be very popular!

The location of a hole is shown by the centre point – where the point of a pair of compasses would be placed, or the centre of a drill bit would start drilling. This is shown with a cross, usually with broken lines. The outside diameter of the circle is then drawn:

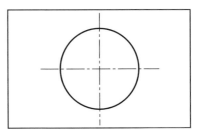

Fig 1.56 Centre line

Dimensions are marked with arrow-headed lines and can be either radius or diameter measurements.

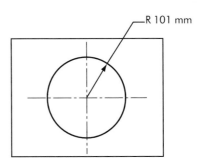

Fig 1.57 Radius

Holes that cannot be seen are shown as dotted circles:

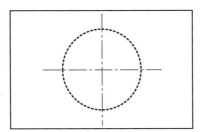

Fig 1.58 A hidden hole

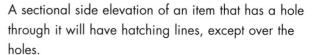

A side elevation will have dotted lines to signify the position and depth of the holes; centre lines will also be needed.

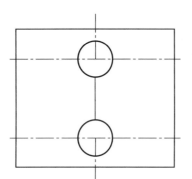

Fig 1.59 Side elevation

A sectional side elevation of an item that has a hole through it will have hatching lines, except over the holes.

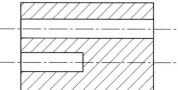

Fig 1.60 A sectional view

ACTIVITIES

Choose a product you can measure that has holes on the surface.

1. Produce an accurate engineering drawing of the surface.

2. Produce an accurate side elevation of the object.

REVISION NOTES

Holes and springs need to be dimensioned in accordance with drawing conventions.

Schematic Diagrams

A schematic diagram shows how things are related to each other. Like the types of engineering drawing already covered they generally use symbols and conventions so that they are easily understood.

There are many different types of schematic diagrams, including:

*   circuit diagrams

*   wiring diagrams

*   pneumatic diagrams

*   hydraulic diagrams.

British Standard PP307 covers many of these.

Electrical and Electronic Systems

The most common way of showing how electrical components fit together is a circuit diagram. These can get very complicated very quickly and so it is

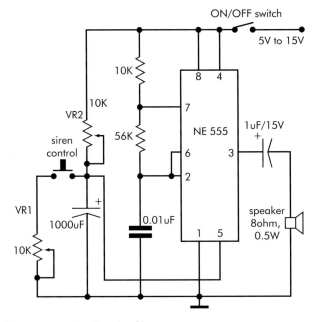

Fig 1.61 A circuit diagram

very important to make sure you draw them correctly. Someone making the circuit, or designing an automated system to build it, needs more information. They will need a diagram showing what goes where, sometimes called a wiring diagram. Both diagrams contain the same information but are used in different ways.

ACTIVITY

Draw a circuit and a wiring diagram for a simple electrical device (e.g. a torch or personal alarm).

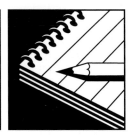

KEY WORD

Schematic   diagram made up of symbols, which gives a plan of how something works, rather than what it actually looks like

— REVISION NOTES —

Many CAD packages have specialist symbols that will help you draw schematics of all kinds. It is essential to include them if your project has circuits of any kind.

If you do your diagrams by hand, follow conventions and label them neatly.

Circuit Diagrams

When drawing a circuit diagram you need to know the following.

- Most use circuit symbols (see below).

- All components should be joined with straight lines (representing wires).

- If you use the correct symbol you do not need to label components.

- All circuits need a source or supply of electrical energy.

- All circuits need some appliance that turns the electrical energy into useful work of some sort (lamp, motor, heater, etc.).

The electrical symbols that you should know are listed below.

Component	Symbol	Description
Resistor		Turns electrical energy into heat
Variable resistor		A resistor whose resistance can be changed (e.g. a dimmer switch)
Thermistor		As the temperature goes up its resistance goes down
Diode		Only lets current pass through in one direction
LED		A light emitting diode
Capacitor		Stores electrical charge
Bulb		Turns electrical energy into light (and heat)
Multi-cell battery		A source of electrical energy
Motor		Turns electrical energy into motion (kinetic energy)
Buzzer		Turns electrical energy into sound

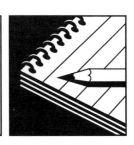

ACTIVITY

Research and find at least four circuit diagrams that use some of the above symbols. Explain what each one does.

Pneumatic and Hydraulic Diagrams

A pneumatic system is one that uses gases and a hydraulic system is one that uses liquids. Other than that they are very similar and so the same symbols and conventions are used to represent them both. The way to tell the difference between the two is that hydraulic systems use arrows that have solid heads but the arrows for pneumatic ones are open. The word fluid can be used to describe both liquids and gases.

hydraulic arrow

pneumatic arrow

Fig 1.62 Hydraulic/pneumatic symbols

The symbols that you should know are:

Valve		Controls flow rate and direction of fluid
Cylinder		Turns fluid pressure into mechanical force
Reservoir		Storage area for fluid (similar to a capacitor in an electrical circuit)
Pipe work		Connects components for fluid to flow down (similar to wires in electrical circuits)
Filter		Restricts flow

ACTIVITY

Research and find at least four pneumatic or hydraulic diagrams that use some of the above symbols. Explain what each one does.

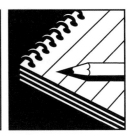

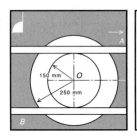
PORTFOLIO NOTES

Follow the rules, use the correct symbols, and include diagrams for all your circuits.

Block and Flow Diagrams

Block diagrams and flow diagrams can look very similar but they are used to show different things.

A flow diagram shows a sequence of events. (They are made up using a fixed set of standard symbols.)

A block diagram shows the relationship between the components or elements of a system. They are useful in the early stages of the design process when you know *what* you want to happen but are not quite sure *how* it will happen.

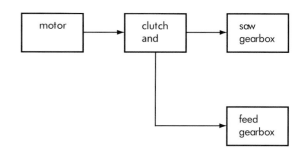

Fig 1.64 A block diagram for a simple bench saw

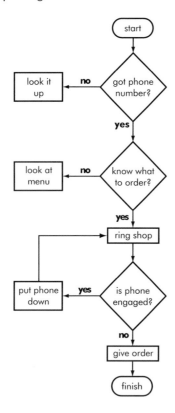

Fig 1.63 A flow diagram to illustrate ordering a pizza

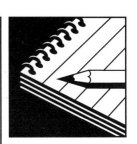

ACTIVITY

1. Draw a block diagram to show the main components in:

   a. a washing machine

   b. a kettle

   c. a pillar drill.

2. Draw a flow diagram to show:

   a. how to make a cup of tea

   b. how a washing machine operates

   c. a central heating system.

KEY WORDS

Block diagram shows the logic of how a system works

Flow diagram shows how parts of the system are related

REVISION NOTES

Flow diagrams can be useful for describing production processes, including quality checks.

Matching Drawing Types to Audience

When you produce a drawing it is important to know who the audience is and what they will be using it for. If I buy a fridge, I do not need an assembly diagram!

The three main types you should consider are:

- working/manufacturing drawings
- service/repair drawings
- assembly drawings.

A working or manufacturing drawing is one that is needed on a regular basis by the engineer who will be manufacturing and/or operating the engineered product. It needs to contain dimensions, materials of construction, tolerances and all production characteristics.

A service/repair drawing is designed for the person who will be servicing and maintaining the product. They may be brought in to make repairs and adjustments. They will need to know both how it fits together and how it can be taken apart.

An assembly drawing is designed for the person putting together the final product. In the case of something like flat pack furniture, this will probably be the end customer. This only needs to contain assembly details and does not need the technical levels featured in the other two.

Fig 1.65 The end customer needs clear and straightforward guidelines

The nature and form of the drawings will also be influenced by whom they are intended for. If you are presenting a final design then you will spend a great deal of time on layout and presentation. You may use graphics, colours and eye-catching images. This is fine for presentation purposes but the service engineer who is fixing the product just needs to know what goes where and how can it be changed or repaired.

The most likely audiences are:

- manufacturing engineers – who build it

- service engineers – who fix it

- technical customers – who use it.

You will need to think about who your drawings are aimed at. A potential customer may be very impressed by a CD-ROM multimedia presentation explaining all the features and information, but a manufacturing engineer is less likely to need this type of information.

Fig 1.66 Think about who will need to use the information

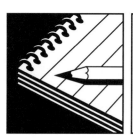

ACTIVITY

Collect examples of each of the types of drawing listed above and write down the key features of each one.

KEY WORDS

Working/manufacturing drawings   help someone to produce a product

Service/repair drawings   help someone to maintain the product

Assembly drawings   will help someone to put the product together

PORTFOLIO NOTES

Think about the many drawing styles you are likely to include in your work. Consider:

- the audience – the people who are going to be using the drawing

- the purpose – what are they going to use it for?

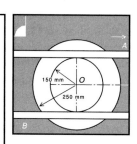

PRESENTING A DESIGN SOLUTION ⬡5

Even when you have decided on a final solution to the original design brief this is not the end of the process. What you must then do is present and justify this information to others, particularly the client. This is likely to consist of lots of different pieces of paper so it is sensible to keep everything together in a design portfolio.

Your final presentation may include:

- the initial design problem and/or client brief

- any research you carried out and the conclusions you have drawn from it

- information about the idea generation process including other possible solutions considered and why they were rejected

- any changes that were made during the process and why they were made

- a clear and well-presented final design solution.

Fig 1.67 Presenting your ideas is an important part of the process

PORTFOLIO NOTES

All through your final design solution you need to show that you have evaluated what you have done. The design is your own work, made up of your choices, but you need to make sure that you justify them.

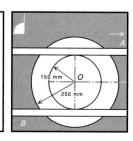

The Initial Design Problem

This section does not have to be very long. It should be short and contain only the key aspects of the problem. This is the reason it is called a design *brief*!

Research

Although you may understand your research (probably because you did it!), you must communicate both what you have found out and how you have used this information in your design work.

Your survey results may be presented in graphs, charts and tables, or perhaps in a list of the most relevant or popular comments made. If you use a good, well-labelled diagram you can show lots of information very clearly.

Whenever you use data from a questionnaire or survey, you must enclose a copy of the questionnaire or survey in your report and also include all the data you collected.

The other research you include really depends on what you have done, but remember research is much more than questionnaires and surveys. Whatever you include, follow the same approach:

- what is the research?
- where is it from?
- how has it influenced or changed the design?

What you are trying to show here is that you have collected enough information to help you make the correct decisions.

Idea Generation

It may be that you have come up with the best possible solution but you have to show that you have considered other possibilities. This does not have to be in great detail but could include information from mind-showering sessions or copies of any idea webs.

You do not need to include a lot of detail if it is not all that relevant. The important thing is to show that you have considered lots of options.

This is the most appropriate place to put information about other possible solutions and why they were rejected.

What you are trying to show here is that you have considered lots of possibilities before picking one solution.

Changes Made During the Design Process

During the process from brief to finished product there will be many changes. Some of these will be as a result of feedback from possible customers during research and sometimes changes will come from the client. You do not need to list every single change but make sure you explain any changes and why they were made.

What you are trying to show here is that in your design work you are able to adapt and change in order to produce the best final outcome.

The Final Design Idea

This is the most important bit; it must contain all the information needed to make the item as well as explaining how it matches the design brief. It is made up of two parts:

- details of your final design idea
- an explanation of how your final design solution meets the client design brief.

Your final design idea should have all relevant details, contain all appropriate drawings and any other relevant information. In this section of the book you have seen how to present information in many different drawings. You will need to decide which ones to use.

Earlier on in your design solution you will have explained in detail the decisions and choices that you have made. What you need here is more like a checklist. Write out each of the key features of the original design brief and then show how the final idea matches this. You will also need to include how it complies with relevant sector standards and legislation (e.g. food guidelines, child safety).

Engineered Products

2

CHAPTER AIMS & INTRODUCTION

This chapter will explore:

- production processes
- teamwork
- product specifications
- production planning
- production methods
- material properties
- quality control
- construction methods
- drives, pneumatics and electrical components
- tools and equipment
- health and safety.

This chapter is the one that helps you get down to the part of this course you probably chose it for – making things. It helps you plan that process carefully, so that your work gets finished on time, to the correct standard, and without anyone getting hurt.

All production processes are different and present different problems, but the types of problem are similar across different operations. It doesn't really matter whether an operator gets long hair caught in a machine for planing wood or cutting metal, they are still going to get hurt, and so the safety considerations aren't really all that different.

Choosing the right tools, machines and processes will improve the chances of a successful product being produced safely. Of course, that does not mean that nothing can go wrong. Practical work often throws up difficulties you were not expecting. If that happens, you need to change your plans and find another way to do the job. That is not a problem, as long as you record what you changed and why.

① THE PRODUCTION PROCESS

In Chapter 1, you have gone through the stages needed to create a product specification (and hopefully created one yourself!). If you were then going on to do the engineering part yourself, there would be no problem in communicating your ideas. This hardly ever happens and so clear communication is vital.

The whole process from the first idea to the finished product is made up of the following stages.

- Start-up – Meetings to develop a feel for the product and identify its requirements and constraints.

- Concept – Development of a rough model (or models) to include all the major features of the final design and prove that it conforms to the requirements.

- Design – Production of the full design and splitting the concept into required parts. Drawing up the final design and checking it conforms to the specification.

- Detailed design – Working out any problems from the earlier stages and producing all required documentation (e.g. drawings).

- Manufacture – All instructions are passed to the manufacturer, so that the product can be produced. Communication continues to take place to iron out any problems.

② PEOPLE, TEAMWORK AND SUCCESS

Although most of the information in this book is about objects, properties and processes, the role that human beings play in this engineering process should not be ignored. An engineering environment is not made up of lots of people doing their own job and never talking to one another. If this were the case nothing would ever get made.

In order for any process to be completed successfully the people involved must work as a team, communicating effectively and working towards the same goal. To become a successful engineer you will need to develop your teamwork skills as well as learn the necessary technical skills. If you have one set of skills without the other you will not be as good an engineer as you could be.

?

REVISION QUESTIONS

1. If you were going to be creating a production plan for a product that one of your friends wanted, what questions would you ask to ensure you produced exactly what he or she had in mind?

2. If someone you did not know gave you the design solution, would this make any difference?

KEY WORD

Communication   sharing ideas with other people, either verbally or in writing

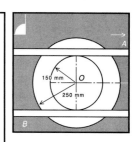

PORTFOLIO NOTES

Remember that your portfolio is what will communicate your design to the moderator who will assess it. You need to make everything clear, easy to follow and easy to find.

A well laid-out table of contents with numbered pages and clearly labelled diagrams will help others to follow your thoughts.

USING A PRODUCT SPECIFICATION ⬡3

Many companies have separate design and engineering sections and so you need to know how to read and use a product specification that is produced by someone else. Having all the right information is vital if your final engineered product is to match the client's requirements.

The key things that should be in a product specification are:

- size, shape, form

- materials, parts and components

- process methods (these may not always be specified)

- quantity required (e.g. single, batch, volume, continuous)

- timescales.

You will have covered most of these in Chapter 1. The information is the same but now the emphasis is different. It has moved from design to engineering. A designer takes an idea and puts it down on paper; an engineer takes a design and turns it into a three-dimensional working object!

Fig 2.1 Agreeing a specification

ACTIVITY

Lots of information is involved in the engineering process. It needs to be communicated between people. List all the possible ways in which any information about the design could be passed on to others.

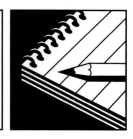

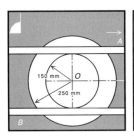

PORTFOLIO NOTES

Make sure your specifications are clear enough for someone else to follow.

4 PRODUCTION PLANNING

When you have a comprehensive product specification, you will then need to turn this into a real object. All the possible aspects of production need to be considered and brought together into something called a production plan. This will include the following things:

- parts and components to be used

- size, shape and form

- materials to be used

- processes, tools, equipment and machinery to be used

- the sequence of production including critical production and quality control points

- production scheduling, including realistic deadlines

- how quality will be checked and inspected

- health and safety factors.

The production plan needs to cover both the processes that need to happen and the order in which they need to happen. Getting things done in the correct order is as important as doing the right thing.

During the design stage you are able to deal with each section step by step. When creating a production plan you need to consider everything in one go. The next few pages go through all the required stages. Many of these have been discussed

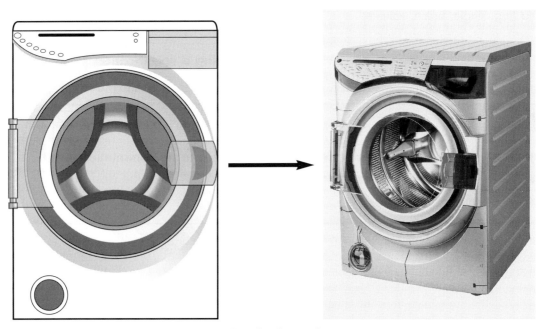

Fig 2.2 From design drawing to finished product

in detail in Chapter 1 but it is worth noting that the emphasis has now shifted from design to engineering (drawing to making!) Some of these things may be specified in the design solution but it is unlikely that all of them will be. The areas not previously covered are detailed in this section. Health and safety is such an important area it is dealt with in its own section later on in this book.

ACTIVITY

Create a list of all the information you would need from a designer in order to create a production plan. Mark and explain any things that are vital.

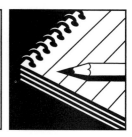

KEY WORD

Production plan   all the stages of production listed in order

PORTFOLIO NOTES

Make sure you have all the information available for your production plan.

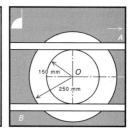

Parts and Components

In order to make any engineered product you must have the specific requirements for all parts. Almost all products will contain more than one part and require some kind of assembly. The component parts that are required may be mechanical or electrical and may be bought in (or out-sourced).

Not all of these parts will be made in the same place by the same people, so it is very important for the information to be clear, unambiguous and consistent across all parts. It is fine to buy in a windscreen, rather than make it in the car plant, but it still needs to fit!

Bought-in parts tend to be standard sizes and shapes. This means that they can be manufactured cheaply in very large quantities and so it is more economical to buy them in. Examples include light bulbs, zips and individual electrical components like switches.

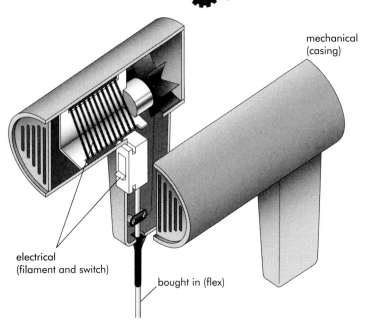

mechanical
(casing)

electrical
(filament and switch)

bought in (flex)

Fig 2.3 A cutaway view of a hairdryer showing its component parts

REVISION QUESTIONS

1. A torch is usually made up of mechanical, electrical and bought-in parts. Draw an exploded diagram of a torch and list all the component parts. For each part state which category it falls into.

2. Pick any two products, write a full parts list for them and write down whether you think the parts are mechanical or electrical and if they are bought-in.

In both questions it will not be possible to know for certain which products are bought in but you can make a sensible guess.

KEY WORDS

Mechanical   operated on or using a machine or mechanism

Engineering   the application of science to manufactured things

Bought-in   components bought from another supplier

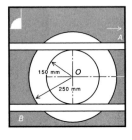

PORTFOLIO NOTES

You will need to produce a parts list for your product. Mark on it any parts that you intend to buy in.

Size, Shape and Form

By the time the design stage is finished all of this information should be finalized. You are expected to include all relevant information in your design solution. There may be some things that you did not consider relevant to your design process but that will need to be known for the production process. Something like a surface finish can easily be forgotten.

There must be enough detail in the specification to ensure that the final engineered product will meet the specifications of the client. Even if you think a piece of information is obvious, it should still be included. Information for all parts and components must be included.

Cutting and Shaping Templates

If products are to be cut or shaped from raw materials, you will need to produce a cutting list or template. This will make it possible to calculate how much raw material is needed and also how much waste will be produced.

Very few materials are totally uniform, so this will add an extra problem when considering the cutting plan. Wood has a grain that makes cutting different depending on the direction in which the cut is made, fabrics may have a weave direction and even metals do not always behave the same way in different directions because of how they were formed.

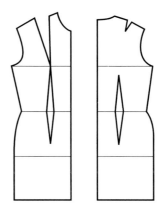

Fig 2.4 This dress pattern is an example of a cutting template

ACTIVITIES

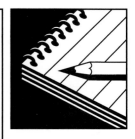

1. Produce a drawing of any product. The drawing must include the dimensions, component parts and materials of construction.

2. Using an A4 sheet of paper and a 15 cm ruler, create two cutting templates. On the first one, get as many ruler shapes on the paper as possible. On the second, arrange the shapes so that as few as possible will fit. In each case make sure there is no space bigger than the size of your ruler left on the paper.

KEY WORD

Template   standard shape to cut from

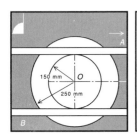

PORTFOLIO NOTES

Include a cutting layout for any parts that will be cut from raw material. If the direction of weave or grain matters, you should specify which way the part should be cut.

The Directional Properties of Materials

To try to help you understand about how materials behave when they are cut or shaped and what effect the direction has, carry out the following experiments. Write a short but detailed report stating what you have found. Diagrams will be a great help.

1. Get a sheet of newspaper.

    a. Tear a piece in half from top to bottom.

    b. Tear a piece in half from side to side.

2. Take a sheet of aluminium foil. Using a sharp point like a compass, scrape it across the foil. Try this in different directions and see if there is any difference.

3. Try to explain why these things happen.

Materials

Choosing materials is very important; you need to consider both their properties and how they can be engineered. This is covered in much more detail later in this section.

Once the materials are chosen, you need to know how much of them you need. When considering the quantities you should consider the following:

- How much wastage there will be.

- In what quantities the raw materials are available.

In some cases it may be cheaper to buy raw materials in bulk to gain a discounted price but there are other things to consider. These include delivery, storage and required environmental conditions (e.g. does it need to be kept warm and dry?).

REVISION QUESTIONS

1. For an engineering process you need 90 kg of lead. It is sold in 10 kg blocks for £18 each or in 50 kg blocks for £80. Which is the cheaper option? Show your calculations.

2. Explain why, if you were dealing with toxic or radioactive chemicals, buying large quantities of raw materials may not always be suitable.

KEY WORD

Quantity   the amount of material you will need to buy

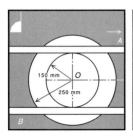

PORTFOLIO NOTES

You will need to think about how your materials would be bought, even if you don't actually have to buy them for your project. You should highlight any that would need special storage arrangements.

Processes, Tools, Equipment and Machinery

Once the engineering process has been chosen, a number of things need to be considered before production can start. Questions might include:

- does the company have appropriate equipment/machinery?

- are the staff adequately trained?

- if not, can this training be done within the company or is outside support needed?

- are there any new health and safety considerations?

- is the correct environment available?

- what knock-on effect will this new production have?

The extent to which these factors need to be considered and acted upon will change depending on the individual situations involved. Below are some brief explanations and possible examples.

Equipment

It most cases the engineering will take place in an already established workshop or factory environment. This means that there will probably already be lots of machinery and tools available. If time and cost were no object, then brand new machinery would be bought for each new production process. The reality is that in many cases any existing machinery available will be used in some way.

Fig 2.5 Do you have the correct equipment?

Training

All the staff involved in the production need to be fully trained. This is both to make sure the products are correctly made and to ensure all the staff are safe. If any new processes are introduced, then the requirements for new training need be considered. In some cases, this might mean that someone from outside the company will need to be brought in. This will be even more likely if brand new processes and machinery are needed.

Fig 2.6 Staff need to be properly trained

Health and Safety

The issues relating to maintaining a safe and healthy working environment are covered later in this book. Even if a thorough health and safety audit has been carried out, any new systems need to be assessed for risk *before* production begins.

Environment

Different regulations and legislation apply depending upon what is being produced. The hygiene requirements for food production are obviously far more stringent than those required in the production of tyres. If new chemicals are to be used, then ventilation systems may need to be checked and updated.

Knock-on Effects

It is almost impossible to say what these might be as they are so dependent upon individual circumstances. What is important is to try and consider all reasonable possibilities. Some examples are:

- internal maintenance/cleaning: if this is carried out by in-house staff, will extra manpower be needed?

- insurance: if risks and circumstances change, the existing policies many not be valid.

- staffing: if this increases, will other facilities need to be changed or upgraded (changing rooms, catering facilities, washrooms, etc.)?

ACTIVITY

If the number of students in your school, college or class was to be doubled, write down all the knock-on effects it would have.

KEY WORDS

Equipment   the tools and machinery that are used to make an object

Training   needed to make sure the staff know what they have to do

Health and safety precautions   rules that are put in place to ensure workers are safe and healthy

Environment   dust extraction, hygiene and other considerations for safety

Knock-on   effect on other parts of the company

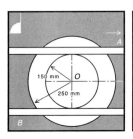

PORTFOLIO NOTES

Make a list of equipment that will be needed to produce your product and any health and safety issues you will need to investigate.

Timing and sequences

In order to make it easier to estimate how long production will take to complete, it is good practice to look at each stage of the production and make an estimate of the time it will take. This information is then used to make an overall estimate.

It is not simply a case of adding up all the individual times, as it is possible that different component parts may be made at the same time in different places. The final assembly cannot begin until all component parts are available to be assembled.

If parts are going to be bought in, then you will also need to consider the time between placing an order for parts and when they actually arrive (sometimes called the lead time).

The cycle time is another important thing to consider in the plan. This is the time between the completion of one item and the completion of the next, which needs to be kept to a minimum. This is usually done by careful planning and sequencing of the production processes. For example, it may be possible to schedule a practical activity to be carried out on one item while another item is drying or baking.

The example below shows how finishes could be applied to a product, the first example has a long cycle time, the second example has a short cycle time, therefore the second example could help improve productivity.

Example 1

Item 1	Spray	Dry	Spray	Dry	Polish	Finished						
Item 2							Spray	Dry	Spray	Dry	Polish	Finished

Example 2

Item 1	Spray	Dry	Spray	Dry	Polish	Finished	
Item 2		Spray	Dry	Spray	Dry	Polish	Finished

This is not always possible, as the different operations may have to be carried out by specialists, or the equipment may be in use for other things, but this method of scheduling must be considered in all manufacturing production where a number of the same items are made.

Critical Production Points and Quality Control Points

The production plan will need to include critical production points and quality control points. Once you have looked at all the processes involved and these points, then you can decide on a realistic schedule. From this it will be possible to set deadlines, which will require a balance between leaving no time in the plan for any problems or errors and having too much slack time.

Critical Production Points

A critical production point is the point in a process where there is the largest likelihood of problems. This is usually when a number of things need to happen before the process can move on. The best illustration of this is the final assembly. This can only happen if all the component parts are finished, ready and available. If they are not, then you cannot begin the assembly process. If something goes wrong when the raw materials are being cut, production cannot continue until the problem has been dealt with.

If you have a clear sequence of what happens when, then it becomes easier to see where such problems may occur. Try to organize things well and take steps to minimize the chances of problems occurring.

Fig 2.7 A clear sequence of events is important

- If a single component is not up to standard the object may need to be taken apart, which may be impossible or very expensive.

- Delays could be created that hold up all production.

- A vital component may not be able to be checked easily (or at all) once assembly has taken place (e.g. it is too late to check that the inside of a golf ball is correct once the outer coating is on it).

- If a large batch is found to be faulty this can prove very expensive.

Quality Control Points

Throughout any engineering process it is important to have a well-organized system that is efficient and cost effective. It is also important to ensure that the final product is up to the specified quality standards.

It is possible to wait to the end of production and then check the finished product but this will have serious consequences if problems have already arisen.

At various points in the production process, quality control checks need to be put in place. The type and nature of these checks is dealt with in much more detail later on in this chapter.

In deciding what is to be checked and when such checks should take place, a balance needs to be struck between the checks slowing the whole production run down and the risk of problems going unnoticed.

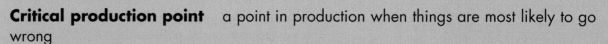

KEY WORDS

Critical production point a point in production when things are most likely to go wrong

Quality control point a stage in the process where the quality of the product is checked

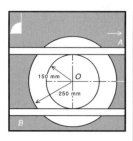

PORTFOLIO NOTES

Work through your production plan stage by stage and list where the critical production points would be.

Work through your production plan stage by stage and list where there should be quality control points.

Quality Checks and Inspections

In any engineering process it is very important that the quality of both the final product and its individual components is checked against fixed standards. Different methods can be used depending upon the type of process involved. These include:

- manual checking

- mechanical checking

- x-ray or ultrasound analysis.

Manual checking can take many different forms. It would be impractical for someone to take each component at each stage of the process and check it by hand. Usually, after a certain number of a product or component have been produced, one is removed and checked. The regularity of this checking would need to be considered. The check may just be a visual one and this could be done via some type of CCTV link. In many cases, the check will need to be more comprehensive.

It is possible to check some things in a mechanized and automated way. If dumbbells are being produced then it is important that the stated mass is correct. It would be ridiculous for someone to take a dumbbell off a conveyor belt and put it on a scale to check. In this case, it would be sensible to build a mass checking system into the production line somewhere.

In other cases, it will only be possible for checking to be done by hand. It is not yet possible to use a computer system to check if soup tastes nice. (Although it is possible to automatically check how much salt it contains!)

Some checks cannot be carried out by a human. X-rays and ultrasound are used to check the internal quality of some materials. These techniques can be used to identify problems within a solid object, such as a void in cast material.

Fig 2.8 Manual checks are made on a newspaper printing press

ACTIVITY

Try to find out in what places x-ray or ultrasonic inspection techniques are used. In each case, find out why it is the most suitable.

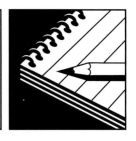

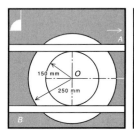

PORTFOLIO NOTES

State what checks and inspections should be carried out at the quality control and critical production points on your plan.

(5) CHOOSING MATERIALS, PARTS AND COMPONENTS

When picking materials for an engineered product, the characteristics of the material must match those in the product specification. It may not always be possible to find a perfect match, so alternatives may need to be used. If this is the case, then the reasons need to be clearly explained.

Many people split all material into metals or non-metals. When more detail is needed we use four groups:

- metals and alloys
- polymers
- ceramics
- composites.

Metals are often described as ferrous or non-ferrous. Ferrite is the old name for iron (iron has the chemical symbol Fe) and so ferrous means 'containing iron'; non-ferrous means that there is no iron present. An alloy is a mixture of two or more metals that are not chemically joined together.

The most common polymers are plastics. It is worth remembering that all plastics are polymers but not all polymers are plastics.

Ceramics have no crystalline structure at all and they are usually very brittle. Glass is a ceramic.

Composites are made up from more than one material mixed together but not chemically joined. Glass-reinforced plastic (GRP) is a composite material.

Fig 2.9 Examples of different materials (l to r): metal, polymer, ceramic and composites

Material Properties

Before looking at the four main types of material listed opposite, we need to know something about the kind of properties we are looking for.

The properties of any material we use can be split up as follows:

- chemical properties
- electrical and magnetic properties
- thermal properties
- mechanical properties.

Each of these will be looked at in more detail over the next few pages.

Chemical Properties

Chemical properties relate to how a material might change chemically in certain circumstances. The type of chemical change that affects metals most frequently is called corrosion. Rusting is the most common form of corrosion and occurs when materials containing iron are attacked by oxygen and water. Rusting is not the only type of corrosion, metals can be affected by all kinds of chemicals.

Non-metals don't corrode in the same way but can still be affected by contact with chemicals or even light. This is called degradation. You will probably have noticed how the colours of some plastic objects become dull when left in strong sunlight. This chemical reaction to the sun's ultra violet light is a type of degradation.

ACTIVITIES

1. Write down how it would be possible to test for some of the chemical properties mentioned above.

2. What steps could be taken to reduce the degradation that would occur in a brightly coloured object that is used in bright sunlight?

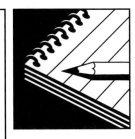

Electrical and Magnetic Properties

If a material allows an electrical current to pass through it, it is known as a conductor. Some materials are better conductors than others, for example gold is a better conductor than copper. In very expensive hi-fi leads, the connectors are made of gold so that the signal is of a better quality. Materials that do not conduct well are called insulators. In many cases the temperature of a material will affect how good a conductor or insulator it is.

Magnetic properties relate to how something behaves in a magnetic field. Some materials will become strongly magnetic when placed in a magnetic field (e.g. iron, nickel and cobalt) and others will be much less affected. Some will keep their magnetism (hard materials) and some will lose it when the magnet is taken away (soft materials).

In Chapter 3, you will see that for many new materials such as superconductors the magnetic properties are very important.

ACTIVITIES

1. Write down how it would be possible to test some of the electrical and magnetic properties mentioned above.

2. Produce a short report on superconductors. Include what they are made from, how they are made and where they are used.

Mechanical Properties

The mechanical properties of a material are the way it behaves when forces are applied to it. Below are some common mechanical properties that engineers consider, although some of these are only really relevant to metals:

- elasticity – how much you can stretch a material so that when you remove the force it still goes back into its original shape

- plasticity – how easy it is to permanently change the shape of a material by applying a force. Plasticity is related to malleability and ductility

- malleability – how well a material can be flattened or stretched into sheets

- ductility – how easy it is to change the shape of a material by applying forces

- tenacity – how well a material can resist being pulled apart. The amount of force needed to pull a material apart is called the ultimate tensile strength.

When changing the shape of an object using force, some materials will change much more easily than others. If an object changes shape easily it is said to be flexible; if it is difficult to reshape then it is said to be rigid. This applies whether the material is elastic or plastic.

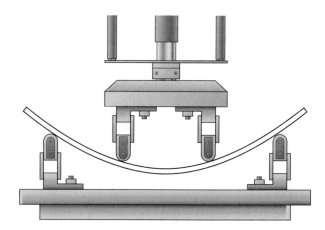

Fig 2.10 Testing mechanical properties of a material

ACTIVITIES

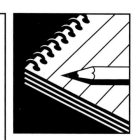

1. Research how each of the properties listed above is tested. You may wish to include diagrams.

2. Copy and complete the table below. For each type of behaviour mark whether it is a mechanical property or not. If it is not a mechanical property write what type of property it is in the next column and an example of a material that behaves that way in the last column. The first one has been done for you.

Behaviour	Mechanical (Yes/No)	If not mechanical, what type	Name of material
conducts electricity	No	Electrical	Copper
can be cut easily			
surface can be scratched			
can be hammered into sheets			
can be lifted with a magnet			
melts on a hot day			
changes colour under ultra violet light			
changes shape when stretched			
dissolves when mixed with acid			
is an insulator of heat			
shatters if dropped			
rusts easily			
expands when heated			

PORTFOLIO NOTES

What properties are important for the materials you have chosen for your product?

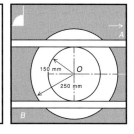

Metals and Alloys

Metals and alloys are used a great deal in engineering products and can have a wide range of properties. It is rare for a metal to be used in its pure form, what usually happens is that metals are mixed together to combine properties. An alloy is a mix of two or more metals. Other, non-metallic chemicals are usually added to improve the properties.

Non-ferrous Metals and their Alloys

There are many useful metals and alloys that do not contain iron. Some of the most commonly used ones are aluminium, zinc, copper, bronze and brass.

Name	Metal or Alloy	Uses and properties
Aluminium	Metal	One of the most abundant metals in the earth's crust. Lower density than steel, so much lighter in construction. Not as strong as steel, so for high performance uses (e.g. aircraft) it needs to be alloyed.
Zinc	Metal	A low melting point (400°C) makes this very useful for casting. Aluminium is sometimes added before casting to increase strength. Used in the auto industry for carburettors, fuel pumps and door handles.
Copper	Metal	A very good conductor (particularly when very pure). Plentiful in supply and relatively cheap. It is flexible, making it suitable for wires, but it lacks strength. Copper oxide can be added to make it stronger.
Bronze	Alloy	Alloy of copper and tin. Very resistant to corrosion and so good for outside statues/sculptures.
Brass	Alloy	Alloy of copper and zinc. The amount of zinc dictates the properties.

Fig 2.11 Table of metals

Ferrous Metals and their Alloys

Man has been using iron for thousands of years. The ore that contains it is widely available throughout the world and, compared to many other metals, it is easy to remove the metal from the ore. Any metallic material containing iron is known as ferrous.

Fig 2.12 Iron has been used in different ways for centuries

Steel is the material most commonly made from iron, and it is made by adding carbon. Sometimes other elements are added to make a more complicated alloy but this is a whole area of engineering in itself. The amount of carbon added dictates what type of steel is made and its properties.

Name	Carbon content %	Common uses
Mild steel	0.05–0.30	Sheets, wires, screws, nails, reinforcement bars, car bodies
Medium carbon steel	0.25–0.5	Shafts, gears, railway wheels, high tensile tubing
High carbon steel	0.55–1.4	Hammers, chisels, high tensile wires, knives, ball bearings

Fig 2.13 Table of common steels

Adding more than 1.4% carbon does not make a big change to the properties. Carbon contents of above 3% do exist. This is called cast iron and is used for machine castings.

ACTIVITIES

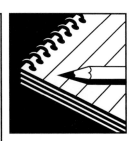

1. Research as many different types of alloy as you can and make notes of their names, composition and uses.

2. There are two types of brass, called Naval brass and Admiralty brass. What is added to the brass and why is this done?

3. Copper oxide is sometimes added to copper to change its properties. What effect does it have?

KEY WORD

Alloy   metallic material made from more than one metal

PORTFOLIO NOTES

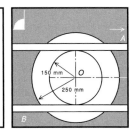

Will you be using any alloys in your product? If so, what has been added to the main metal and why?

Polymers

A polymer is a material that is made up of long chains of chemicals joined together. You will not need to understand the chemistry of them, but the way these chains behave affects the properties of materials. The two main polymer types you will need to know about are thermosetting plastics and thermoplastic polymers.

Thermosetting Plastics (Thermosets)

Thermosets are usually made available in powdered or granular form. When they are heated and moulded they form a rigid material. This happens because, when heated, the long chains form very strong bonds that cannot be broken by heating.

When a thermosetting plastic is heated, it does not melt. The bonds between the chains are so strong that the material will decompose and burn. Thermosetting plastics are used for things like saucepan handles.

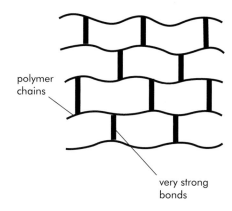

Fig 2.14 A thermoset polymer chain

ACTIVITIES

1. Collect as many different types of plastic you can. Try to identify them, state their key properties and where they are used.

2. Research what recycling facilities there are in your local area for plastics and other materials.

Thermoplastic Polymers

Thermoplastic polymers behave in a different way. When they are heated they can be shaped, but this time the bonds between the chains are much weaker. This means that if they are heated again, the bonds weaken and allow the material to be reshaped. They are usually more flexible and not as strong as thermosetting plastics.

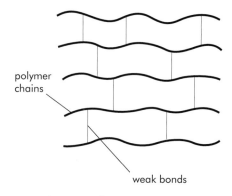

Fig 2.15 A thermoplastic polymer chain

Name	Type	Properties and uses
Epoxy resin	Thermoset	Very strong, used as an adhesive. A very good insulator so it is used to coat and protect electrical components.
Polyester	Thermoset	Usually used with other material to make polyester laminates. These are resistant to heat, water, chemicals and impact. Used in bathroom fittings, water tanks, safety helmets. Most commonly used in glass- and carbon-reinforced plastic (GRP, CRP).
Acrylics	Thermoplastic	Very good optical properties, can be shaped by heat easily and a good insulator. 'Perspex' and 'Plexiglas' are acrylics.
Polythene	Thermoplastic	Not very strong but very easy to mould. Waterproof properties and flexibility make it useful in sheeting for foundations in buildings.
Polyvinyl chloride (PVC)	Thermoplastic	Tough and rubbery, easy to manipulate with good electrical properties. Used as an insulator coating for electrical cables.

Fig 2.16 Table of common polymers

KEY WORDS

Thermoset   a plastic that remains rigid when heated

Thermoplastic   a plastic that melts when heated

PORTFOLIO NOTES

Identify any plastics you intend to use. Are they thermosetting or thermoplastic?

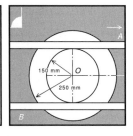

Ceramics

The term 'ceramic' includes many materials other than pottery and porcelain, and includes some of the earliest materials made by man. Ceramic is fast becoming the material of the future. It is now used in some of the most sophisticated superconductor technology. This technology could lead to us all using levitating trains and having much more powerful computers.

Fig 2.17 Levitating trains use ceramic technology

Ceramics have an amorphous structure, which means that they have no clear shape or crystalline structure. They tend to be very strong in compression (squashing forces) but weak in tension (stretching forces). They are very good insulators of heat and electricity. The circular red things you see on power lines are ceramic. The tiles designed to protect the space shuttle on re-entry are ceramic.

Because of their properties, ceramics have a wide and varied use in engineering. Examples include:

- grinding tools
- cutting tools
- seals
- bearings
- heat and electrical insulation.

Glass is a common and widely used ceramic material that is particularly useful because of its optical properties. These are governed both by the way glass is made and by its components. Some common glasses and their uses are as follows.

- Soda glass (sometimes known as soda-lime-silica glass): the most commonly used type of glass; its electrical properties are not as good as other types.

- Lead glass (sometimes called flint-glass): of a much higher quality; it is softer than other glasses so it is easier to cut and grind; used for optical purposes.

- Borosilicate glass: has a high resistance and low thermal expansion and is therefore used for thermal and electrical applications.

Fig 2.18 Ceramics are used for both electrical and heat insulation

ACTIVITY

Write a short information sheet on ceramics. Explain how their uses have changed, from the time man discovered how to make them through to their use in today's ultra modern technology.

Composites

A composite is a material that is made up by combining two or more materials in order to merge their properties. As engineering has become more sophisticated, the requirements for new materials has increased. This has meant that using a single material is not always sufficient. By combining materials, we can combine their properties and create a 'new' material. The materials are not chemically joined, they just 'share' their materials.

The development and advancement of composite material technology has meant that many industries including aerospace, motoring and construction, have been able to go bigger, higher and faster than ever before!

Fig 2.19 Formula One racing cars are an example of composite technology

Concrete

Concrete is very strong in compression but weak in tension. In order to give concrete more strength, steel rods are inserted into the concrete as it sets. Steel is very good in tension so the reinforced concrete has the compressive strength of concrete and the tensile strength of steel.

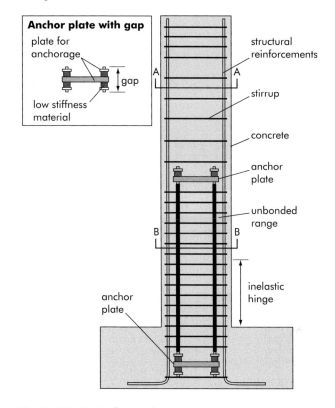

Fig 2.20 Reinforced concrete

ACTIVITY

Research pre-stressed reinforced concrete. How is it different from conventional reinforced concrete? Where is it used and how is it produced?

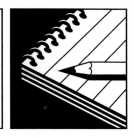

Glass- or Carbon-Reinforced Plastic (GRP/CRP)

GRP and CRP follow exactly the same principles as those used in reinforced concrete but apply them to different materials with different properties.

Very thin glass and carbon fibres are remarkably strong in tension (stretching) but need to be held in a fixed shape in order to be useful. They are put into an epoxy resin (called a matrix), which holds the fibres in place.

Composites combine the properties of the two and have:

- high strength

- low density

- good rigidity.

They are used in making the bodywork for formula one racing cars, helicopter blades, fishing rods and aeroplane wings.

Fig 2.21 Aeroplane wings are made from GRP/CRP

ACTIVITY

Research other composites that are used. Include as much detail as you can about the properties involved.

KEY WORDS

Ceramics   hard, brittle materials made from non-metallic materials heated to high temperatures

Composites   materials made from two or more materials, which are not chemically joined

GRP   plastic reinforced with strands of glass, an example of a composite

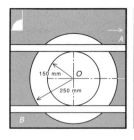

PORTFOLIO NOTES

Are you using any ceramic or composite materials? What properties of the material made it useful for your product?

Sintered Metals

Some metals have a very high melting point; this means that shaping them can be very expensive. Sintering is a process that allows the shaping of metals at relatively low temperatures. It is also possible to mix in different metals and other materials during the sintering process in order to combine properties.

Tungsten melts at over 3400°C but can be sintered at around 1500°C. This difference has a dramatic effect on the costs involved. The process involves using metal in powdered form and is described in outline in the following steps.

1. Metals are ground up into powders.

2. The powder particles are sorted by size.

3. The correct grade of powder is selected.

4. If required, different powders are thoroughly mixed.

5. The powder is compacted into a die.

6. The temperature is increased until a coherent product is formed.

This process can save a great deal of money by turning raw materials into finished products which cost less and use less sophisticated machinery.

Bi-metal Strips

Two metals that expand at different rates are joined together. When they are heated, one piece of metal will expand more than the other and so the strip will bend. The bigger the difference between the two metals the more it will bend.

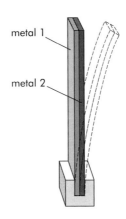

metal 1

metal 2

Fig 2.22 Bi-metallic strip

Although this may seem a little crude, a bi-metallic strip will behave in a very fixed and predictable way. They are used in thermostats and as thermometers to make or break electrical circuits controlling temperature. The strip is usually made into a coil which can have a dial attached. If it is fixed at one end, the more the temperature rises the more the coil will twist.

ACTIVITY

If all metals expand when heated, try to answer the following puzzle. If a metal washer is heated up, does the hole in the middle get smaller, larger or stay the same? This is not as easy as it looks!

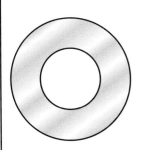

HEAT

????

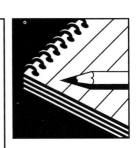

6 MECHANICAL COMPONENTS

You will come across many mechanical components in engineering; it would take a lifetime to learn about all the ones that are available. Below is a list of the main ones you will come across in your work. You are expected to learn these.

Where fixings are involved in a product, the material of the fixing has to be carefully considered. The fixing must be able to withstand the same environmental factors as the components that it is joining (e.g. corrosion, high or low temperatures).

Nuts and Bolts

This is a widely used joining method. It has the advantage that it is not a permanent join.

The diameter of the bolt will be chosen according to the load that it is expected to bear. It is also designed so that there is an equal chance of failing by the head pulling off, the thread stripping or the threaded part of the shank failing.

There is a tendency for nuts to become loose, especially if they are subject to vibration. It is possible to use special locking devices to eliminate this problem.

Fig 2.23 Nut and bolt

Screws

When using screws, the length, the screw thread and screw head all have to be considered. The thread affects the strength of the fixing, but not all types of

thread are appropriate for all materials. The head type can be slightly less functional and more aesthetic. In most cases the final choice will be a balance between cost, strength, how easy it is to tighten or loosen, and appearance.

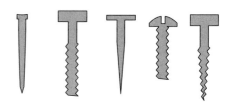

Fig 2.24 Types of screws

Rivets

Rivets are used to make permanent joints. The rivet has a pre-formed head on one end. The other end is passed through a hole in the materials to be joined and that end is then formed (or squeezed) so that the join is fixed.

The size of the rivet is dictated by the thickness of the materials to be joined. They are good at resisting shear (sideways) forces but not at resisting tensile (stretching) forces. The material of the rivet must be reasonably malleable so that the head can be formed.

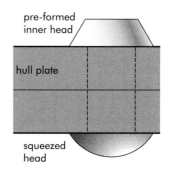

pre-formed inner head

hull plate

squeezed head

Fig 2.25 A rivet

ACTIVITIES

1. Research different types of screw threads and screw heads, and where and when they are used.

2. Find about the differences between snap, pan head, flat, mushroom-headed and pop rivets.

Springs

A spring is basically a coil and is also known as a helix. Springs can be found in many engineered products. They are often used to allow movement within a device, such as a valve.

Generally a spring is used to force two components apart. A second fixing, such as a nut and bolt, is then used to force the same two components together. This arrangement means that the device is kept in equilibrium, until something happens to tip the balance in favour of one or the other.

It is possible to make a spring on a lathe but this can be dangerous and the speed at which the lathe is used must be very slow indeed.

PORTFOLIO NOTES

1. Draw up a table to show a definition and use of each of the following:

   • nuts and bolts

   • screws

   • rivets

   • springs.

2. Whatever fixing methods you use for your product, you will need to specify them carefully in terms of material, size and detail, such as thread size and type. This information usually goes on the parts list.

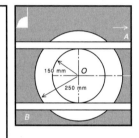

Pins and Clips

Shafts need to rotate, which presents a problem when it comes to fixings. Shafts are often fixed with a mechanical fixing like a bolt that is then held in place with a pin or clip. This is a removable fixing that allows maintenance and cleaning to be carried out easily. The clip fits into a groove prepared for it and the pin fits into a hole.

Pins

There are several different types available. A split pin passes through the hole and then the ends are bent back to hold it in place. The principle is just the same as a paper fastener. Other pins stay in place by virtue of their design or properties.

Fig 2.26 A split pin

Clips

These can be designed to fit either internally or externally. Circlips are the most common type of clip used but more recently wire rings have become popular.

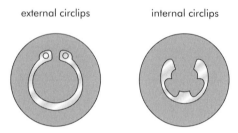

external circlips   internal circlips

Fig 2.27 Circlips

Keys

If you have one object that rotates, like a shaft, and you wish to attach another object to it, like a pulley, you would use a key. The purpose of the key is to keep the two objects in a fixed relative position.

A slot is milled along the length of the shaft and a corresponding slot is milled in the pulley. These holes are called keyways. A key is placed in the keyway on the shaft and then it is put in place. The pulley is then put over the other half of the key and the pair is secured. The key is intended to maintain position only, not to transfer any forces, but in many cases it will do so.

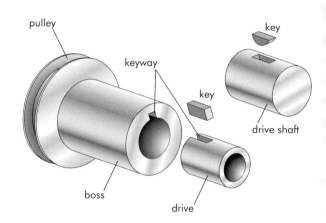

pulley   key
keyway
key
boss   drive   drive shaft

Fig 2.28 Keys and keyways

This is not a permanent fixing and so makes maintenance and cleaning easier. If one part fails then it can be replaced, whereas if the objects were permanently joined, the whole unit would need to be replaced.

In most cases the key will be rectangular in shape and its dimensions will depend upon the diameter of the shaft.

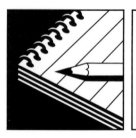

ACTIVITY

Find common objects that use the mechanical components discussed above. Draw a diagram for each one showing how the fixing is arranged and how it works.

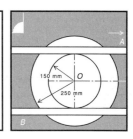

PORTFOLIO NOTES

Clips and keys will need to be shown clearly in detail drawings.

Drive Mechanisms

Many machines use some kind of rotary motion that needs to be transferred to the place where it is needed. This is done by a rotary system or drive mechanism. The most commonly used systems are gears, pulleys and belts. In all cases, they can change the size and direction of the force.

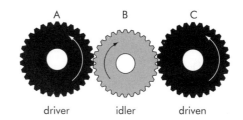

Fig 2.30 Gears

Gears

A hand whisk has a set of gears, called a gear train, that mesh together. These transfer the direction of the force – your hands do not move in the same direction as the whisk blades. They also make the whisk blades turn faster than your hands could do and this is caused by the gear ratio.

Fig 2.29 Two examples of gear trains

If two or more sets of gear trains are available and joined together in some way, then this is called a gearbox.

Each individual gear in a car (1, 2, 3, reverse, etc.) is a gear train in itself, as it uses a number of gear wheels to take the force from the cylinder to the wheels. When all the gear trains in a car are connected together, this is the gearbox.

If you have only two gears in a chain, an anticlockwise rotation will drive a clockwise rotation. If the direction needs to be the same for both input and output then a third, idler, gear is used. The only function of the idler is to change direction, hence its name.

The gear ratio for a gear system is calculated as follows:

$$\text{Gear Ratio} = \frac{\text{Number of teeth on driven gear}}{\text{Number of teeth on driving gear}}$$

For example, I have one gear with 15 teeth and it is driving another one with 30 teeth. What is the gear ratio?

$$\text{Gear Ratio} = \frac{30}{15} = 2 \text{ or } 2{:}1$$

The drive shaft has therefore to turn twice in order for the driven shaft to turn once.

ACTIVITY

Copy and complete the following table. The first calculation has been done for you.

Teeth on driver gear	Teeth on driven gear	Gear ratio
15	30	2 :1
250	250	
36	9	
200	100	
4000	10,000	
10		1:3
	100	4 :1
30	15	

KEY WORDS

Gear train   series of connecting gears

Gear box   series of connecting gear trains

Gear ratio   the number of teeth on a driving gear

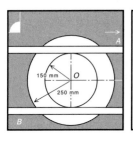

PORTFOLIO NOTES

Exploded drawings are often a good way of showing the details of a gear train. If the gears you use are standard bought-in ones, include them in your parts list. If you are actually making the gears yourself, include drawings of them.

Belts and Pulleys

Gears are not the only way in which a turning force can be transferred; belts and pulleys can also be used. Gears transfer torques very well, but belt and pulley systems are much lighter and quieter when performing the same function.

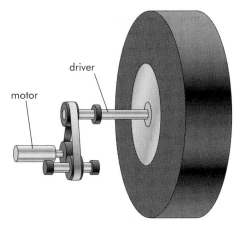

Fig 2.31 Belt/pulley system

Belts and pulleys require friction between the drive belt and the rotating shafts in order to work. If the frictional force needs to be increased, the shape and composition of the belt can be changed.

The tension in the belt needs to be maintained for it to work efficiently and to avoid slipping. In order to do this a jockey pulley is used. This can easily be moved to maintain tension.

The velocity ratio for a belt and pulley system is very similar to the gear ratio and is worked out in the same way.

The driver pulley rotates the belt, which transfers the rotating motion to the driven pulley. The diameters of the pulleys affect the difference in the speeds of rotation of the two shafts.

$$\text{Velocity Ratio} = \frac{\text{Diameter of driven pulley}}{\text{Diameter of driving pulley}}$$

REVISION QUESTIONS

Why do you use a chain and sprockets on a bicycle and not pulleys and belts?

KEY WORDS

Pulley a wheel fixed on to a shaft

Drive belt a belt linking the pulleys on two shafts

Velocity ratio the relationship between the speeds of two pulleys, calculated by the diameter of the driving pulley

Jockey pulley adjustable pulley used to maintain tension in a drive belt

PORTFOLIO NOTES

Describe and justify any pulley systems you use, and show your velocity ratio calculations.

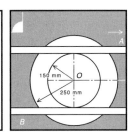

7 ELECTRICAL AND ELECTRONIC COMPONENTS

A number of basic electrical components are used to make straightforward circuits. As the circuits become more complicated then it is likely that pre-assembled/integrated circuits will be used. Below is a list of the main components you will come across. You need to learn these.

Component	Circuit symbol	Property/behaviour
Resistor		Turns electrical energy into heat; used to control the flow of electrical current in a circuit.
Capacitor		Stores electrical charge; can be used as timers in circuits as they charge and discharge in a fixed amount of time; used in conversion of AC to DC.
Diode		Only lets current pass through in one direction; these are used in circuits that change AC to DC such as rectifiers.
LED		A light emitting diode; used as indicator lights to show that something is turned on.
Bulb		Turns electrical energy into light (and heat).
Wire		Connects components, allowing current to flow.
Cable	N/A	Wire or wires, each protected by a sheath of insulation; may also have an outer sheath (e.g. mains power cable).
Insulator	N/A	Resists the flow of electric current (has extremely high resistance); used to protect from electric shock.
Multi-cell battery		A source of electrical energy produced from chemical energy.
Motor		Turns electrical energy into motion (kinetic energy).

Component	Circuit symbol	Property/behaviour
Buzzer		Turns electrical energy into sound; used in alarms.
Variable resistor		A resistor whose resistance can be changed; used when current needs to be controlled, as in a dimmer switch.
Thermistor		As the temperature goes up, its resistance goes down.
Transistor		A component that can control the current flow in a circuit based on an input signal; used in high speed switching and amplifying circuits.
Integrated circuit	N/A	All the components needed for a particular circuit; are pre-attached to a silicon slice and connected correctly to each other; used for amplifiers, counters, processors and oscillators.

Fig 2.32 Table of electrical components

ACTIVITY

Research and find at least four electrical circuit diagrams that use some of the above components. Explain what each one does.

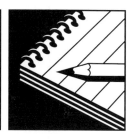

PORTFOLIO NOTES

Include circuit diagrams in your portfolio, and include the components used in your parts list.

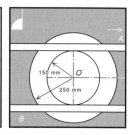

PNEUMATIC AND HYDRAULIC COMPONENTS ⬡8

Pneumatic and hydraulic systems use symbols for drawing purposes. The most common ones with an explanation of what they do are listed. The way to tell the difference between the two is that hydraulic systems use arrows that have solid heads but the pneumatic ones are open. The word 'fluid' can be used to describe both liquids and gases. Hydraulic systems use liquids; pneumatic systems use gases.

 hydraulic arrow

 pneumatic arrow

Fig 2.33 Symbols for hydraulic and pneumatic systems

The components that you should know are:

Component	Standard symbol
Directional control valve	
Flow control valve	
Cylinder	
Reservoir	
Filter	

Fig 2.34 Table of hydraulic and pneumatic components

Valve

There are two main types of valve. A directional control valve (DCV) will control flow lines in a system. The flow lines or paths are called ways. This means that a three-way valve has three paths through it. A common example of this is the controls for a heating system in a car.

Fig 2.35 A valve regulates the car's heating system

A flow control valve is used to regulate the amount of fluid passing through it, called the flow rate. A tap is a common type of flow control valve. The more you open it, the faster the flow becomes, which increases the volume of water per second.

When working with fluids, the volume of flow per second is often more relevant than its speed. It does not really matter whether the water comes out fast or slowly, what does matter is how long it takes before your bath is full and you can hop in! Speed measures the rate at which something moves; volume of flow is the amount of liquid transferred over a period of time.

Cylinder

A cylinder is used to turn fluid pressure into mechanical force. Cylinders are often found inside internal combustion engines. Here the pressure increase caused by the expansion of a burning fuel is converted into a mechanical force and used to move the vehicle forward.

Reservoir

A reservoir is a storage area for fluids (similar to a capacitor in electrical circuits). In many central heating and hot water systems, there is a reservoir of hot water kept in the loft.

Filter

A filter restricts flow by only allowing particles below a given size to pass through it. You may be used to the idea of a filter separating solids from liquids, but it has many other functions. A filter can be designed to separate one fluid from another. Breathing apparatus has filters that are used to remove smoke particles before they enter the lungs, but filters can also be designed to prevent certain gases passing through them.

Fig 2.36 Protective breathing apparatus use filters to block dangerous particles

ACTIVITY

Research and find at least three pneumatic or hydraulic system diagrams that use some of the above components. Explain what each one does.

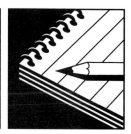

KEY WORD

Hydraulic system   a control system that uses a liquid such as oil

PORTFOLIO NOTES

Design a control system using a gas such as air. Include flow diagrams for any hydraulic or pneumatic circuits you use.

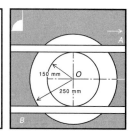

THE PROPERTIES AND CHARACTERISTICS OF MATERIALS ⑨

When choosing a material, its properties and characteristics will be important, as they control how the material can be formed and used. It is possible to change those characteristics in a variety of ways. Some of the most common ones are described in this section.

Shaping and Forming

Process	Notes and comments
Hammering	Hitting an object with another solid object in order to change its shape. The repeated force will change the shape and the chemical structure of the material. This will only work if the material behaves plastically.
Casting	Heating a material above its melting point and pouring the molten material into a mould.
Forging	Heating a metal to soften it and then deform it to the required shape. When cooled, it will retain its new shape.
Forming	Deforming a metal without the introduction of heat, e.g. using a press.
Bending	Changing the curvature of a material. This is particularly relevant to rods and pipes. When pipes are bent, specialist tools are needed so that they do not crease or crush.
Coiling	Making wires into springs. This can be done on a lathe but care is needed.

Fig 2.37 Table showing ways to shape materials

Fig 2.38 A pipe bending tool

Chemical and Heat Treatments

Materials can be treated by the effect of heat or by the effect of contact with other chemicals. In both cases, the aim is to modify the material so that its properties change.

Chemical treatment is a term that covers a number of different procedures. The most common types are electroplating and etching.

- Electroplating is a process where a conducting material is placed in a chemical solution and then current is passed through the solution. A thin layer of plating material is deposited on the object being electroplated.

- Etching is a process where selected areas of a material are dissolved by chemical action. This is most commonly used in the production of printed circuit boards.

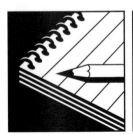

ACTIVITY

Research more details about the use of etching in the production of PCBs. What chemicals are involved and what safety precautions are required?

Heat treatment is a process used on metals. It involves the following stages:

- heating a metal up to a certain temperature

- keeping the metal at a set temperature for a certain amount of time

- cooling the metal at a certain rate.

The times and temperatures involved will change the structures in different ways. The three most common heat treatments are annealing, tempering and hardening.

Surface Finishing

It may be that the properties most suitable for manufacture do not match the properties needed for the surface of the product. This may be for aesthetic reasons (it looks nicer) or practical reasons (I do not want it to rust). It is possible to coat the surface to satisfy both requirements. The most common surface finishing is painting. Chrome plating is also used. Metal goods can be coated with a layer of plastic.

ACTIVITY

Write a short report on the different types of paint that are available and in what ways they differ. A field trip to the local DIY store should give you all the information you will need.

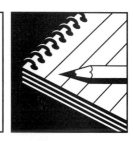

KEY WORDS

Heat treatment   the act of heating and cooling a metal to change its chemical structure and properties

Chemical treatment   the act of applying chemicals to the surface of a product

Surface finishing   adding an external surface to a component to improve it in some way

PORTFOLIO NOTES

What techniques will you use to modify the properties of your raw materials to make them suitable for the job you need them to do? Make sure you explain the purpose of each process. Are you using a painted finish, for instance, because it looks good or to stop a ferrous component rusting?

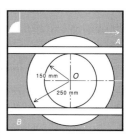

Ease of Handling

This is particularly relevant for objects that will be handled regularly, such as taps, light switches, tool handles and mobile phones. It is easy to increase or enhance the function of something like a mobile phone by cramming in all the latest technology, but if it does not feel comfortable to hold or use then no one will buy it.

The material you may want to use must satisfy both the functional and the aesthetic criteria of the product.

Fig 2.39 Choosing the right materials is important for both function and aesthetics

Cost

The cost of materials is very important and must be built into any plan. This is even more important for materials that fluctuate in price, such as precious metals, oil and food products.

It is advisable to compare the costs of a number of alternative materials and components. The cost of an item must take into account both its quality and its availability. There is no point in using a supplier that advertises very low prices if the goods you want are never actually in stock.

Bulk buying may reduce the cost and provide 'economies of scale' – the ability to buy at a low price per unit – but it may also create extra costs such as storage. A furniture maker with a very small workshop may find it cheaper to order wood in 20 m sections and saw it himself, but if his workshop is too small to accommodate large sections of timber or too much time is involved in cutting it, he may have to spend more by buying pre-sawn timber.

Availability

You may find the right product at the right price, only to be told that it will take six months to deliver and it is only available in 20 m sheets that you cannot cut! The size, form and regularity of supply must be considered.

Manufacturers rely on a regular and consistent supply of material from their material suppliers; if this chain fails, then production ceases and money is lost. Therefore, when selecting a material with the right properties for the product, the supply chain also needs to be reliable.

The form in which material is delivered to the manufacturer is also important for storage, transportation costs and any initial preparation that may be necessary before the manufacturing process can begin. Wood comes in planks, sheets, blocks, etc. The right choice will depend on what processes are going to be used.

Fig 2.40 This material is delivered in rolls

Wool is delivered in bales

ACTIVITY

Sheet steel is only available in certain thicknesses. Find out what these are and why this is the case.

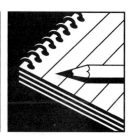

KEY WORDS

Economies of scale   buying large quantities in order to reduce the price of each product

Availability   the material is in stock and can be delivered when you need it

PORTFOLIO NOTES

In what form would you buy your raw materials if you were producing your product on a commercial scale? Can you buy in material that is close to the size and form you want, to reduce production time? If so, would it be much more expensive?

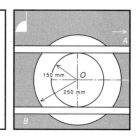

USING ENGINEERING PROCESSES ⟨10⟩

Using an engineering process to produce a component is an area where you will need to have practical experience as well as factual knowledge. The processes below are the ones that you need to know about. In each case you should be able to explain their importance both functionally and aesthetically.

Methods of Constructing Products

Until relatively recently, engineering processes and materials were often considered to be specific to a particular trade. Metal was used in the metal industry and woodworkers used wood. However, this has now changed; there are many similarities between the tools and processes used with different materials. The same basic principles are used to work the majority of materials and the tools that are used share similar or common features.

When making products, the processes fall into a number of categories:

- material removal
- joining and assembly
- treatment processes
- surface finishing.

Material removal is sometimes called shaping by wastage – the material is cut to size or shape, or excess material is removed using various methods, such as:

- cutting – using hand tools or machines
- filing or grinding
- drilling, milling, turning.

For each of these methods the material starts off larger than needed and ends up smaller and lighter.

> All you ever make is sawdust – the piece you want was always there before you started.
>
> *W. J. K. Gough*

Joining and assembly is concerned with the methods of fixing component parts together. This may be permanent or temporary. Temporary fixings can be removed later if required.

Fig 2.41 Traditionally, woodworkers have only worked with wood – now this is changing

Permanent fixings include welding and glueing. Temporary fixings include mechanical fixings – nails, screws, nuts and bolts, hinges – and dry joints – twisted wires, fielded panels.

Treatment processes generally involve applying either heat, or a chemical, or both, to a material to change it in some way. For example:

- heating a steel bar to enable it to be forged or bent
- etching copper on a printed circuit board
- plating metals by anodizing.

Surface finishing means applying a desired finish to a material to protect it, or improve its appearance in some way, for example:

- painting
- polishing
- lacquering
- enamelling.

ACTIVITIES

1. Take a product you have studied. Make a list of the processes that were used in making it.

2. Break the list of processes down into:

   • material removal

   • joining and assembly

   • treatment processes

   • surface finishing.

PORTFOLIO NOTES

Make a similar list for your product.

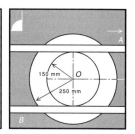

TOOLS AND EQUIPMENT ⬡11

For any process you will need to select and use the appropriate tools and equipment. This is not always just a question of which saw to pick up, it may include the possibility of using computer-aided manufacture (CAM).

The things you will need to consider when making your choices include:

• availability

• cost

• ease of handling

• properties of materials and components.

The type of tool needed for any process will depend on the material, for example, a plaster cast is removed with a vibrosaw to prevent damage to the skin; cheese is cut with a thin wire; wood needs a saw with large teeth, but metal needs one with small teeth. It will also depend upon circumstances. It may be easier to use a heavy-duty grinding machine for a process but if the work has to be done in a small confined space then this may not be the correct choice.

Fig 2.42 Always select the appropriate tool – you don't need a sledgehammer to break a nut!

Cost and Maintenance of Tools

When buying tools for home DIY use, many people buy the cheapest ones. This is not a cost-effective approach in industry. Cheap tools will have lower quality standards than more expensive ones and may turn out to be a false economy.

Carpentry tools are a carpenter's livelihood, and he or she cannot afford for them not to work. They need to be cleaned and maintained regularly and stored correctly. Regular maintenance will lengthen the useful life of most tools.

If members of the workforce are sharing tools and equipment then they all have a responsibility of care.

This not just a case of good working practice, there are also health and safety issues involved. Damaged tools that are left lying around a workshop can present considerable risks.

Fig 2.43 Don't leave things lying around!

ACTIVITY

Write a brief maintenance sheet for four common workshop tools. This should include the correct way to store them, any cleaning that needs to be done and how often. Any parts that require regular checking, sharpening or replacing should also be included.

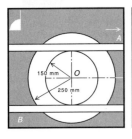

PORTFOLIO NOTES

Your risk assessment should include checking that the tools are in good working order before use.

HAND TOOLS ⬡12

Hand tools can be classified into different categories:

- marking out, measuring and testing tools
- holding tools
- driving tools
- cutting tools.

Rules for Marking Out

- Always work from a true edge or datum.

- On wood, use a pencil for construction lines and a marking knife for the lines to be cut.

- On plastic, use a wax crayon or felt tip for construction lines and a scriber or marking knife for cutting lines.

- On metal, use a scriber to mark construction lines and a dot punch to mark cutting lines (a number of dots 5 mm or so apart).

- Use a marking gauge to mark lines parallel to an edge on wood.

- Use odd leg callipers to mark a line parallel to an edge on metal or plastic.

- Use a tri-square to mark lines at right angles to a true edge.

- Always mark out clearly and accurately, making sure that waste is clearly identified.

	Wood	Plastics	Metal
Angle plate		Yes	Yes
Callipers		Yes	Yes
Centre punch		Yes	Yes
Compasses	Yes	Yes	
Dividers		Yes	Yes
Dot punch		Yes	Yes
Drill gauge	Yes	Yes	Yes
Engineer's square		Yes	Yes
Marking gauge	Yes	Yes	
Marking knife	Yes	Yes	
Micrometer		Yes	Yes
Mitre square	Yes	Yes	
Pencil	Yes		Yes
Rule and straight edge	Yes	Yes	Yes
Scriber		Yes	Yes
Sliding bevel	Yes	Yes	Yes
Surface plate and gauge		Yes	Yes
Templates	Yes	Yes	Yes
Tri-square	Yes	Yes	
Vee block		Yes	Yes
Wax crayon/felt tip pen		Yes	
Wire and sheet gauge		Yes	Yes

Fig 2.44 Table of marking out tools

ACTIVITIES

1. Take a piece of scrap material and make a true edge – plane timber and file plastic or metal.

2. Using the true edge, mark out a regular shape, such as a square or rectangle.

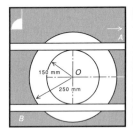

PORTFOLIO NOTES

Identify the marking out tools you will need. You may want to include a photograph of the marked out work before you cut it.

	Wood	Plastic	Metal
Anvil			Yes
Bench holdfast	Yes	Yes	
Bench hook	Yes	Yes	
Bench stop	Yes		
Bench vice	Yes	Yes	
Engineer's vice		Yes	Yes
Folding bars			Yes
G cramp	Yes	Yes	Yes
Hand vice		Yes	Yes
Mitre box	Yes		
Mole wrench		Yes	Yes
Sash cramp	Yes	Yes	Yes
Smith's tongs			Yes
Stakes			Yes
Toolmaker's clamp		Yes	

Fig 2.45 Table of holding tools

Fig 2.46 A vice is often the best way to hold material in place

Rules for Holding Material

- Ensure that the material is free from loose coverings.

- Hold the material with something that is made from the same material or softer; protect softer materials with pads of waste material.

- Do not over-tighten holding equipment.

- Ensure that material is held firmly and securely before carrying out any work.

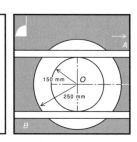

PORTFOLIO NOTES

The way you secure the material you are working on affects the safety of the process. Include details of it in your risk assessments.

	Wood	Plastic	Metal
Ball pein hammer		Yes	Yes
Carpenter's brace	Yes		
Chest brace		Yes	Yes
Claw hammer	Yes		
Electric hand drill	Yes	Yes	Yes
Forging aids			Yes
Hand drill	Yes	Yes	Yes
Pincers	Yes		
Pliers			Yes
Rawhide mallet			Yes
Rivet snap			Yes
Rubber mallet			Yes
Screwdriver	Yes	Yes	Yes
Sledge hammer			Yes
Soft-faced hammer			Yes
Spanner			Yes
Tap/die holder		Yes	Yes
Warrington hammer	Yes		
Wooden mallet	Yes		

Fig 2.47 Table of driving tools

Rules for Driving Tools

- Never hit a softer material with a harder mallet or hammer – the surface will be damaged.
- Do not strike small components with large-faced implements.
- Always use the correct size and pattern of screwdriver.

		Wood	Plastic	Metal
Chisels	Bevel-edged	Yes		
	Cold			Yes
	Firmer	Yes		
	Gouges	Yes		
	Mortise	Yes		
Drills	Taps and dies		Yes	Yes
	Auger	Yes		
	Centre bit	Yes		
	Centre drill	Yes	Yes	Yes
	Countersink	Yes	Yes	Yes
	Expansive bit	Yes		
	Flat bit	Yes		
	Hole saw	Yes	Yes	Yes
	Tank cutter		Yes	Yes
	Twist drills	Yes	Yes	Yes
Files	Needle files		Yes	Yes
	Rasps	Yes	Yes	Yes
Knives	Hot wire		Yes	
	Laminate cutters		Yes	
	Trimming knife	Yes	Yes	
Planes	Planes	Yes		
Portable electrical devices	Angle grinder			Yes
	Circular saw	Yes	Yes	Yes
	Jigsaw	Yes	Yes	Yes
	Polisher		Yes	Yes
	Router	Yes	Yes	
	Sander	Yes	Yes	
Saws	Abrafile	Yes	Yes	Yes
	Bow saw	Yes		
	Coping saw	Yes	Yes	
	Crosscut	Yes		
	Dovetail	Yes	Yes	
	Hacksaw		Yes	Yes
	Junior hacksaw		Yes	Yes
	Pad saw	Yes	Yes	Yes
	Panel	Yes		
	Piercing saw		Yes	Yes
	Rip	Yes		
	Tenon	Yes	Yes	
Snips	Bench shears			Yes
	Curved snips			Yes
	Jeweller's snips			Yes
	Scissors		Yes	
	Straight snips			Yes
	Wire cutter			Yes
Specialist milled tooth files			Yes	Yes
Surforms		Yes	Yes	

Fig 2.48 Table of cutting tools

Rules for Cutting Materials

- When using cutting tools by hand, always try to work vertically or horizontally.

- Effort used to work the tool should always be directed away from the body.

- Space should be made for using the tool safely, without catching on other materials.

- Tools must be kept sharp.

Fig 2.49 Care should be taken when using any cutting tools

ACTIVITIES

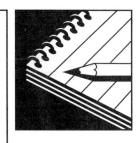

1. Taking the piece of material you marked out earlier, select the appropriate tools to remove the waste material.

2. Check the shape you have cut. Is it a regular shape, as you intended, if not why not?

PORTFOLIO NOTES

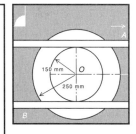

Cutting processes are extremely dangerous if not carried out properly. Anything that is capable of cutting wood, metal or plastic is also capable of cutting you. Your risk assessment should cover the tools to be used, the way the work is to be secured and any protective clothing needed.

Using Hand Tools

It is important to understand the hazards that can be associated with the use of hand tools. Hand tools can cause severe injury when used improperly.

Using the right tool and using it properly reduces the potential of an accidental injury. It is also important to maintain tools properly; sharp tools need less force to apply them and are therefore less dangerous than blunt ones.

Hand tools are so simple and basic that we often don't think about them in terms of safety. The majority of accidents in a workshop are with hand tools. In industry, machines are so well guarded that it is very difficult to get near any moving parts, therefore it is difficult to come to any harm. However, by the very nature of hand tools, they cannot be guarded, and so people are therefore more likely to come to harm using them.

Each tool is designed for a specific purpose and should only be used to carry out that purpose.

Hammers should not be used to drive in screws; scribers should not be used to bore holes through timber!

Fig 2.50 Protective face and headgear should be worn when using a handtool

Hand tools are designed for their jobs. Accidents usually happen because of the way they are used. Many injuries result from the improper use of hand tools or the lack of proper personal protective equipment. These include:

- broken bones and bruises from tools that slip or fall

- eye injuries from flying chips and cutting tools that slip

- cuts and even amputations from saws or cutting tools

- severe lacerations and potential infection due to dirty tools

- puncture wounds from flying chips or sharp objects.

Whenever a tool is used, there is a hazard. The user must be properly trained and the correct safety or protection equipment must be available. The user must also be focused on what they are doing.

Some of the protective equipment generally available includes:

- safety glasses – tinted or clear depending on inside or outside work

- safety goggles – chemical-resistant or impact-resistant

- face shields – chemical- or impact-resistant

- hardhat or headgear

- gloves – work, chemical, water, heat/cold, puncture-resistant

- respiratory protection – dust mask, self-contained breathing apparatus (SCBA)

- foot protection – steel-toed, puncture-resistant, chemical-resistant.

Workers should always follow these simple rules:

- choose the right tool for the job

- make sure you understand how to use it correctly

- make sure it is in good working condition

- think about what you are going to do before you do it

- wear protective equipment if the task requires it.

ACTIVITIES

1. List the safety aspects you considered before marking out and cutting the shape from the waste material used in the previous section.

2. List the protective equipment you used and explain why you used it.

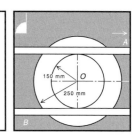
PORTFOLIO NOTES

Hand tool operations should be included in your risk assessments.

APPROPRIATE USE OF MACHINES ⬡13

In many engineering environments there may be a choice of different machines or pieces of equipment that could be used to carry out particular operations. It is important that the right machine is chosen for the job.

When selecting machines, consideration should be given to:

- the operation
- the worker
- the set-up
- the workspace
- access to the workspace
- the work height
- the correct tool size.

All of these factors affect the efficiency of the machine and its capacity to do the task required.

Every machine must be examined before each use. Employees responsible for the machine, and workers using machines on the job, should be taught how to check equipment so that worn or damaged items can be repaired or removed from service.

Machines that should be removed from service include:

- lathes/milling machines with cracked or worn jaws
- pillar drills with broken bits, faulty chucks or weak, loose or broken handles
- power hammers with loose heads or a chipped striking surface
- dull band and radial saw blades
- frayed cords, broken plugs or switches, or damaged extension cords.

It is essential that even fully-trained experts still follow the manufacturer's recommendations and manuals when using machines and always use the proper protective equipment. Many people think they are good enough at a particular operation not to need to wear protective equipment – accidents don't seem to care how skilled a person is!

KEY WORD

Selecting   choosing a machine to use, taking all the relevant factors into account

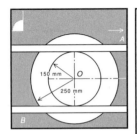

PORTFOLIO NOTES

Carry out an inspection of all of the equipment you will use to manufacture your product.

1. Record your findings in a log.

2. Report anything you feel needs to be acted upon.

(14) CORRECT USE OF HAND AND POWER TOOLS

The safest tool is the one made for the job, used as designed.

At work, tools are only to be used in the ways intended by the manufacturer. 'Creative' use, or misuse, of tools often leads to injury or a damaged product. For example, an adjustable wrench can tighten a nut, but it would be better to use a box-end wrench or a socket wrench of the proper size. Ergonomically-designed tools, of course, provide the best fit for the employee and the job.

To avoid a hazard when using hand and power tools, select the right tool. When selecting a tool, the same considerations apply as when selecting a machine. Tools must always be kept in good condition, following the manufacturer's guidelines. Every tool should be examined before it is used. Employees should be taught how to check equipment so that worn or damaged items can be repaired or removed from service.

Tools that should be removed from service include:

* wrenches with cracked or worn jaws

* screwdrivers with broken bits (points) or broken handles

* hammers with a loose head or chipped striking surface

* dull saws, saw blades, and drill bits.

Follow all labels and manufacturer's recommendations when using tools, and always use the proper personal protective equipment. Store tools in a safe place, either in the work area or in a common toolbox or rack. Proper storage will mean the tools are less likely to be damaged.

Employees' Responsibilities

Employees should be responsible for:

* selecting the proper tool for the job

* checking to ensure that the tool is in good working condition

* using the tool correctly, including proper use of personal protective equipment such as safety spectacles

* cleaning and storing the tool properly.

Supervisors' Responsibilities

Supervisors should be responsible for:

* explaining the job to the employee

* providing adequate training for any tools used

- supplying protective equipment and explaining the need to use it

- ensuring that the correct tools are available

- arranging regular maintenance checks.

ACTIVITIES

1. Ensure all tools are stored safely when not in use – produce a set of rules for using tools.

2. Design a tool storage area that protects the tools and makes it easy to check if any are missing.

KEY WORD

Recommendations   most tools will come with manufacturer's instructions about use, maintenance and storage

---

PORTFOLIO NOTES

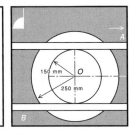

Produce a maintenance cycle for the tools you will be using for the manufacture of your product.

CARRYING OUT ENGINEERING PROCESSES ⬡15

Using a Centre Lathe

The centre lathe is used to turn accurate cylindrical or conical shapes. It can also be used to flatten the surface of a material, or to drill and bore holes.

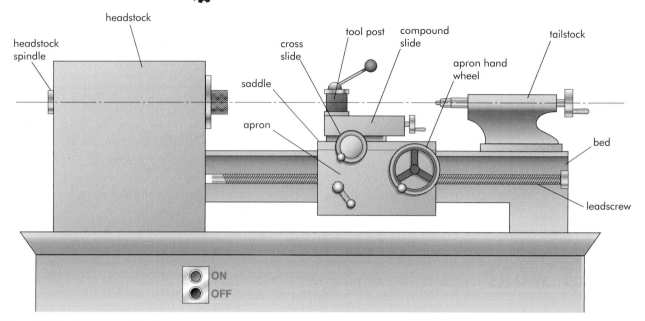

Fig 2.51 A central lathe

The most common operations carried out on a lathe are reducing the diameter of a piece of material and facing off a piece of material.

Material is usually held in a chuck. This can either be a three-jaw chuck, used for holding round or hexagonal material, or a four-jaw chuck, in which all the jaws move independently thereby enabling awkward shapes to be held firmly.

The gearbox, usually located in the headstock, drives the chuck. This means that the material, or work-piece, rotates and the tool used to shape the material remains stationary.

If the work-piece is long enough, the tailstock should be used to support it as it rotates. This stops the material being pushed off-centre by the tool.

Before using a lathe, it is important that it is set at the correct speed and that the correct tool is available. To set the correct speed, a number of facts need to be considered.

- What material is to be turned?

- What operation is to be carried out?

- Is the operation required to produce a high quality finish?

Basically, the softer the material being turned, the faster the lathe speed: a 10 cm work piece of aluminium or nylon can be turned at 10 000 rpm, whereas a 10 cm piece of cast iron should not be turned at above 1000 rpm.

The operation also has a lot to do with what speed is selected.

- If the work-piece is being knurled – a heavy textured finish, often found on metal tool handles – the lathe should be run at a third of the normal turning speed.

- Drilling should be done at three-quarters of the turning speed.

- Parting – cutting the material using a parting tool – at a third of the turning speed.

- If the turning operation is required to produce a high quality finish, the speed should be less than that for a normal cut.

The depth of cut controls how quickly the tool is fed into the work-piece. Setting a large depth of cut removes material quickly, but the finish is likely to be poor. Soft materials can stand a much greater depth of cut than hard ones, which are likely to damage the tool if fed too quickly. These materials will need to be cut in stages.

CAD/CAM systems will automatically select the speed and depth of cut for an operation depending on the material being used.

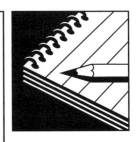

ACTIVITIES

1. Produce a table listing the cutting speeds available on a centre lathe that you could use to make your product.

2. Add the materials you could use to the table, linking them to the appropriate speeds.

3. Make a note of any alterations that you will need to make to the lathe speed as you work through the processes to make your product.

KEY WORDS

Chuck   device for holding a piece of work

Speed   rotation speed of the lathe

Depth of cut   how deeply the tool is set to cut into the piece of work

PORTFOLIO NOTES

Include details of how you intend to carry out any turning on your product.

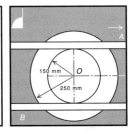

Drilling

The purpose of drilling is to make new holes or enlarge existing ones. The drill bit is rotated at a fixed speed and introduced to the material to be drilled. The drill is then moved into the material at an appropriate rate until the hole is created. This is a very important technique as so many engineered products require holes.

The friction created between the drill and the material is considerable. When wood is being drilled, it is not uncommon to see smoke coming from the wood and scorch marks on the inside of the hole. This is hardly surprising considering you can make fire just by rubbing two sticks together very quickly.

Fig 2.52 A pillar drill and a hand-held drill

The speed of rotation and the type of drill bit will depend both on the material being drilled and the diameter of hole required. In many home drill sets you will see that there are different drill bits for wood, metal and masonry as well as different settings on the drill depending on which bit you are using.

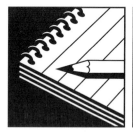

ACTIVITY

Make notes on the differences between drill bits used for wood, metal and masonry. Good clear diagrams will help you here. You will be able to see how the shapes are different but will probably need to look up the materials that they are made of.

KEY WORD

Drill bit   the spiral tool that fits into the chuck of the drill and does the cutting

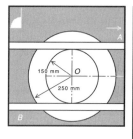

PORTFOLIO NOTES

Drilling operations need to be carefully considered when planning your production, because you need to make sure you can secure the work-piece safely before you drill it. This will be easier at some stages of the product than others.

If you have lots of holes to drill in identical products, it may well be worth making a jig or a template to help you to do them quickly and accurately.

Milling

This is a process applied to metals, using a multipoint tool. More than one cutting edge is used at one time, speeding up the process. The tool is usually fixed and the material is fed past it. The milling can be horizontal or vertical depending upon the arrangement of the tool.

Although the forces involved are high, it is still possible to obtain a good quality machined surface. Milling is usually used where flat, level surfaces are required.

Grinding

This process removes material by abrasive action upon the surface. A grinding wheel is the most common way of performing this action. It is similar to milling as each abrasive particle acts as a miniature tool. Although each individual abrasion cannot be controlled and regulated in the same way as in milling, the interactions are so small that it is possible to get a very good quality flat surface.

Fig 2.53 Using a grinding wheel

Shaping and Manipulating

Heat is not always introduced when changing the shape of materials. Hammering, forming and bending can all take place at room temperature.

Hammering

When repeatedly hitting the material with a hammer or similar tool, two things happen. One is that the object changes shape and this deformation is permanent (plastic deformation). The other is that the crystal structure of the metal is altered.

Heat treatment can change the properties of the grains in metal. Likewise if you repeatedly hit a piece

of metal you will cause the grains to become damaged and distorted. This will mean that they become smaller, resulting in a harder but more brittle material.

Forming

This is another way of deforming a metal without the introduction of heat. The tools and techniques used will depend on both the original shape and the desired new one.

If a material is rolled the effect on the crystals being rolled in one direction – i.e. forward and backward – would be different to the effect on those being rolled the other way – i.e. from side to side. Common types of forming include:

*   rolling
*   corrugating
*   stretching
*   wire drawing
*   stretch forming.

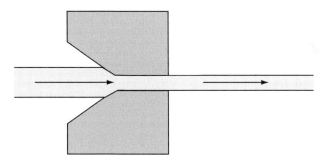

Fig 2.54 An example of wire drawing

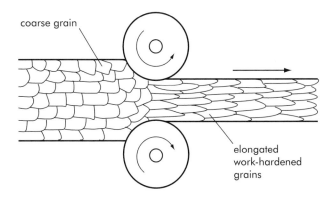

coarse grain

elongated work-hardened grains

Fig 2.55 Sheet rolling

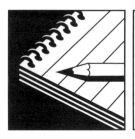

ACTIVITY

Explain the difference between 'cold working' and 'hot working'. Describe the advantages and disadvantages of each. Make some reference to the chemical structures and how they change during these processes.

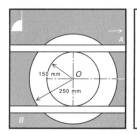

PORTFOLIO NOTES

If you use any of these techniques, you should describe how and why.

Bending

Bending changes the curvature of a material. This has its most common use in rod and pipe work. When bending a pipe, it is important to bear in mind that the inside diameter should be as unaffected as possible. If the pipe is bent too quickly then it is likely to squash in at the bend and may acquire creases. How far round the pipe is bent will also affect this.

Special tools are available for bending and, in the case of pipe bending, it is possible to put a flexible guide inside the pipe in order to minimize the danger of problems occurring. This is done using a lubricated spring. It is not considered good practice to bend pipes over your knee!

Joining and Assembly

Crimping

This is when materials are squashed together. A variety of specialist tools are available. Multi-strand wire is sometimes crimped at the end in order to make a better contact. This process is quicker and cheaper than soldering.

Fig 2.56 Using a crimper

Soldering

Soldering is the process of a making a sound electrical and mechanical joint between metals by joining them with a soft solder. This is an alloy of lead and tin with a low melting point. The joint is heated to the correct temperature by using a soldering iron. It is mainly used for making electrical connections in circuits and joins between pipes in plumbing.

Although the materials that are to be joined do not melt, the tin reacts chemically with the surfaces being joined, making a solid bond.

Although solder melts at around 190°C, the tip of the soldering iron can reach 250°C and so care needs to be taken when soldering.

Fig 2.57 Soldering

Brazing

The process and principles are exactly the same as in soldering. The difference is that brass is used as the solder rather than tin. Brass has a much higher melting point than tin and so the temperatures involved are higher. In brazing, the flux is not melted by electrical heating because of the high temperatures involved (up to 1000°C). A brazing torch is used instead.

Welding

Welding is a process by which two metals are melted and joined together. When they cool, they have effectively become a single piece of metal. It is common to use a *filler metal* during welding, which has a similar composition to the two pieces being joined. Welding is a difficult skill to master. Though it can be more costly, it has largely replaced riveting.

The temperatures involved are very high. These can be achieved either by using an oxy-acetylene flame or an electric arc. Arc welding is when a very high current is used to generate the heat. The light given off is very intense and protective clothing and filtering eye protection is needed during the operation.

Flux is required for arc welding but is not needed for oxy-acetylene welding. Coating the electrode provides the flux.

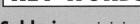

KEY WORDS

Soldering   joining metal parts using a low melting point alloy based on lead and tin

Brazing   joining metal materials with an alloy containing brass

Welding   joining two pieces of metal by melting and fusing their surfaces

REVISION NOTES

Soldering, welding and brazing are all potentially hazardous operations. You must make sure you fully understand what is involved, carry out risk assessments, and then follow them carefully.

Adhesion

Adhesion is the sticking together of materials. The type of adhesive will depend upon exactly how the bond is made. Sometimes the adhesive fills the spaces between materials and sometimes it actually bonds chemically with both sides (see Figure 2.58 overleaf).

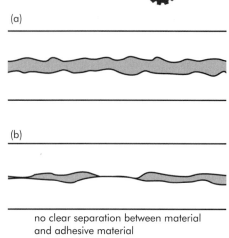

(a)

(b)

no clear separation between material
and adhesive material

Fig 2.58 Surface adhesion

The strength of the join will depend upon the state of the surfaces before adhesion. They need to be clean and free of dust and dirt. If very smooth surfaces or shiny metals are to be joined then they need to be sanded or 'keyed' before joining. This will create a greater surface area available for contact and adhesion because a rough surface can have a greater surface area than a smooth one.

Adhesion can be particularly difficult if dissimilar materials, e.g. metals and plastics are going to be joined.

ACTIVITIES

1. There are some two-part adhesives available. How are they different from one-part adhesives?

2. Find out the difference between cohesion and adhesion.

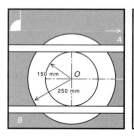

PORTFOLIO NOTES

You may need to test a variety of adhesives before finding the best one for your purposes. Many adhesives need to be stored carefully, and will be subject to the Control of Substances Hazardous to Health (COSHH) regulations.

Wiring

In some cases the easiest thing to do is to tie one thing to another. On a racing motorbike that is subject to intense vibrations it may be practical to tie many of the components in place. This can help to prevent spillages of oil and fuel at high temperatures.

Wiring is also used in welding. If the wire is of a similar composition to that of the filler it can be used to tie the components in place. When heat is applied the wire will melt into the joint.

Heat Treatments

Heat treatment involves the following stages:

• heating a metal up to a certain temperature

• keeping it there for a certain amount of time

• cooling it a certain rate.

The three most common types of heat treatment are annealing, hardening and tempering. Each of these is designed to change the chemical structure of the metal and thus change its properties. You will not need to know much about the structure of metals but the following ideas will help you understand how heat treatments works.

- Metals have a grained structure.
- These grains are usually very small but can sometimes be seen.
- The size of these grains affects the properties of the metal.
- Large grains mean a softer, more ductile metal.
- Small grains mean a stronger but more brittle metal.
- Heat treatment changes the grains and so changes the properties.

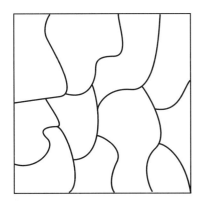

Fig 2.59 Grain structure

In each type of heat treatment there is always a trade-off between properties. If you want a metal to be stronger it will be less ductile. If you want a softer metal it will become weaker.

Annealing

Annealing is when you heat a metal up slowly to a certain temperature and then keep it at that temperature for a certain period of time. The material is then cooled at a specific rate. By controlling the cooling, large crystals are allowed to form. This makes the metal softer and more ductile.

before heating

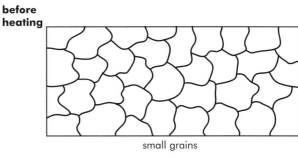

small grains

after heating

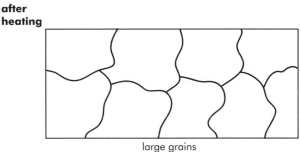

large grains

Fig 2.60 Changes to grain structure

Hardening

The amount of carbon present in steel affects how hard it is, but so does the speed at which it is cooled. In hardening, the material is heated up and then cooled very rapidly (quenched). This means that there is not enough time for large grains to form. The metal grains are small and arranged randomly, making the metal harder and stronger but less ductile.

The metal is often quenched in oil. This means that there are a number of very important safety factors to consider.

ACTIVITY

Research the safety factors involved in a range of metal treatments.

Tempering

This is a follow-on process from hardening. After a material has been quench-hardened it is not always ready for immediate use. A second heat treatment process can be introduced, which is called tempering. This results in the loss of some of the hardness but makes it tougher and less brittle.

Tempering is similar to hardening in process. The metal is heated to a fixed temperature and then quenched. As the temperature of the metal increases, the colour of the metal's glow changes from a pale straw to a bright blue. This colour is used as a guide to the temperature.

Fig 2.61 Tempering changes a metal's glow

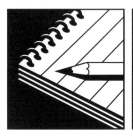

ACTIVITY

Find out more about tempering. Include the temperatures, the colours and what components are tempered at each temperature.

KEY WORDS

Heat treatment heating and cooling metals in specific ways to change their properties

Annealing slow heating and cooling to create large grain size, ductile material

Hardening rapid heating, followed by rapid cooling to create small grain size, hard, brittle material

Tempering a process to make hardened materials less brittle

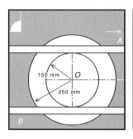

PORTFOLIO NOTES

Include details of any heat treatment you intend to carry out and what material properties you intend to improve by doing it.

Chemical Treatments

Etching

Etching is a process in which selected areas of a material are dissolved by chemical action. It is used for creating a surface finish or as a decorative feature. This can be seen on items as diverse as teapots and trumpets. The areas that are in contact with the chemicals and the time of contact must be carefully controlled in order to obtain the desired effect.

Etching is commonly used in the production of printed circuit boards. The chemicals and ultraviolet light involved can be dangerous, so all health and safety procedures must be followed.

Copper printed circuit boards are made in the following way.

1. Copper is used to coat a plastic or fibreglass board.

2. A negative image of the board is made, often on computer.

3. Parts of this negative image are coated with a chemical that is resistant to the etching solution.

4. The board is exposed to the etching solution.

5. When cleaned, dried and finished, the result is a printed circuit board that is ready for use.

Fig 2.62 Etched glass

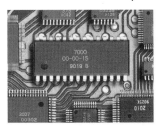

Fig 2.63 A printed circuit board

ACTIVITY

Investigate the chemicals involved, and the safety precautions relating to, etching.

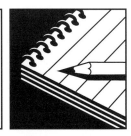

PORTFOLIO NOTES

If you are producing your own PCBs, include the design layout in your portfolio.

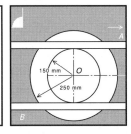

Plating

This involves covering the surface of one material with a thin layer of another and is done for three main reasons:

- decoration
- hardening
- corrosion resistance.

Electroplating is the most common type of plating. It is a process whereby a conducting material is placed in a chemical solution and then current is passed through the solution. Metal particles are displaced from the solution and deposited on the object to be plated.

By changing the concentration of the solution and the strength of the current, this thickness can be varied. This is most commonly used for putting expensive decorative finishes on cheaper materials or making the surfaces of materials more resistant to corrosion.

The diagram below shows the principle of electroplating. On an industrial level the science is the same but the scale much bigger.

Surface Finishing

Most surfaces are finished either by polishing or coating.

Polishing

This involves rubbing the surface with a non-abrasive cloth. As well as creating a shinier surface on metals, it can also make them more resistant to weathering. The polishing makes the grains on the surface smaller and less porous and also reduces friction.

Coating

This is commonly done in three ways: painting, covering in plastic and laminating. In all cases, a new layer is placed upon the existing surface. The choice of coating will be partly aesthetic and partly down to the ease of application. The purpose of the new layer can be to provide a barrier to reduce the effects of weathering or environmental attack, or it could be applied for aesthetic reasons.

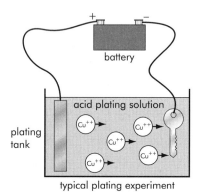

typical plating experiment

Fig 2.64 Electroplating

ACTIVITIES

1. Search around your home and find as many objects as possible that have a coating of some sort. In each case describe the coating and why it has been used.

2. Some coatings are permanent and some need maintenance or replacement. Give some examples of each.

QUALITY CONTROL TECHNIQUES ⬡16

In order to ensure that the final product matches both the design specification and the relevant regulations and legislation, the quality needs to be checked during the production. These checks will have to be performed on individual components as well as on the final product.

The purpose of quality control is to ensure, in a cost efficient manner, that the product sent to customers meets their specifications. Inspecting every product is costly and could not be justified in most cases, but delivering faulty components is not going to please the client, and must be avoided.

Fig 2.65 Quality control is important

Statistical quality control is the process of inspecting a sufficient number of products from a given lot to ensure a specified quality level. This check could be done randomly or after a certain number of the product or product parts have been produced.

Automated checking devices using sensors are often used to check a number of features but manual checking is still common in certain circumstances.

At the end of the production line, more complex products are manually checked and approved. Cars, TVs and many electrical goods are checked in this way. On many of these types of goods you may find a sticker marked 'good', 'ok' or 'checked'.

Important features in a specification include:

- dimensions, tolerances and fit
- finish
- performance
- overall quality.

Some of these, such as the dimensions, the fit and the performance will be dependent upon each other and have been covered already in this book. Some of the others are explained in more detail below.

Overall Quality

This is a combination of all the other factors listed. In order for a final product to be considered a quality object, all individual quality standards must be met. There is no point having a perfectly-designed bicycle, with brilliantly-machined parts, to exact specification and a perfect finish, if the wheels fall off!

ACTIVITY

Write down the stages involved in making a pencil. List all the quality control checks that you think would be needed and when they should happen.

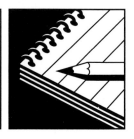

KEY WORD

Quality control systems put in place to make sure the quality of the product does not fall below an acceptable level; this does not necessarily mean every component will be perfect

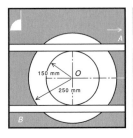

PORTFOLIO NOTES

Include a list of quality control checks to be made on your product. Include spaces to record the results, then you can fill in what happens when the test takes place.

Dimensions, Tolerances and Fit

In the design solution and production plan, the dimensions and their acceptable variations (dimensional tolerances) should be clearly stated.

Some of these dimensions may be related to the aesthetic appearance of the object and some may be related to mechanical function. Whatever the reason these sizes are the ones needed for the final product.

The quality control system needs to be at least partly independent of the manufacturing system. For example, if the same equipment is used to mark the size of an item out before it is cut and is then used to check that it has been cut correctly, there is a real possibility of errors occurring. If the calibration on the cutting side fails or is incorrect, it will not be picked up because the checking system will contain exactly the same error.

Fig 2.66 Some dimensions are related to aesthetics and function

In Chapter 1, tolerances were discussed in relation to dimensions but tolerance can relate to any variable property, from how long a screw thread needs to be to how many matches there should be in a box.

The tolerance specifies how much any quantity can vary from the specification. If you are running a production run of one million parts it is almost impossible to get everything exactly the same size, but it is possible to produce them within a tolerable range, so that they still work. In an engine, the tolerance between components will be microscopic, whereas in a pre-built section of a house it may be much larger.

ENGINEERED PRODUCTS

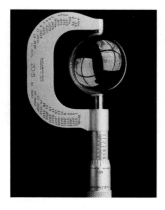

The tolerances on parts in a jet engine are probably as close to consistent accuracy as humans can get, because the parts are exposed to tremendous forces and any failure would be disastrous. Very close tolerance levels in any product will be expensive and consequently will put the price up. However, generally speaking, the closer the tolerance the higher the quality of the product is likely to be.

Fig 2.67 Using a micrometer to test tolerance

ACTIVITY

Every day we use objects that require close tolerances in order to work. The top of a pen has a very low tolerance otherwise it would not stay on. What else that you use every day has been engineered to close tolerance levels? List and explain as many examples as possible.

PORTFOLIO NOTES

Include tolerances on your technical drawings and parts lists. Think carefully about how tight they need to be. Very small tolerances can be difficult to achieve and may not really be necessary.

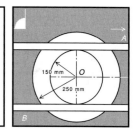

Performance

Many products will be designed to have a long lifespan and therefore part of the quality control work assesses whether that product will last. This is usually done by some kind of mechanical automated test although there are other options available, including computer simulations.

All products need to be fit for their purpose. A brand new car may look and feel like a high quality product. It may even drive very well indeed, but it needs to checked to make sure that it will not fall apart after a year.

If something has a relatively short lifespan and is used continuously then it is reasonably straightforward to test. If it requires a longer lifespan, then the conditions of its use need to recreated. Some examples would include:

* clothes going through a washing and drying cycle continuously to look for fabric wear

* a tennis racquet being automatically hit with the force of a tennis ball, thousands of times

* the effect of someone heavy getting up from and sitting down on a chair, thousands of times.

Some situations cannot be recreated so easily. If you were creating a storage system for nuclear waste, which needed to be secure for tens of thousands of years, then other approaches, including computer modelling would be needed.

Finish

The finish of a product is very important to us, and is linked to the aesthetics. We like to see smooth edges, sharp (or beautifully rounded) corners and accurate printing, all of which help to attract us to a particular product.

The term 'finish' can mean many different things. The finish is not always smooth, it may be textured, ribbed, matt, gloss and so on. The required finish will be described in the product specification.

Fig 2.68 Finish is all-important

The way in which the finish is checked may depend upon whether specialized machine finishing is required or if the finishing is done by hand. If there is a measurable quantity such as a friction coefficient, then this can be measured, possibly automatically. In some cases, human inspection will be the only way to check.

ACTIVITY

Pick four different objects that have a specific surface finish. Describe the object and the finish. Try to research or suggest how the finish was obtained and why that finish was decided on.

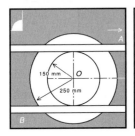

PORTFOLIO NOTES

Try to finish your product to a high standard, and include photographs of it in your portfolio.

17 HEALTH AND SAFETY

In any place of work there will be a number of regulations and laws that apply to health and safety issues. Their job is to make sure that the working environment is safe and hazard free. The best known of these laws is the Health and Safety at Work Act (1974). This contains many individual regulations and includes those relating to substances considered hazardous to health, called COSHH.

Fig 2.69 The Health and Safety Executive and COSHH help to protect workers

In an engineering workplace there will be health and safety issues relating to the use of materials, components, tools and equipment. You will be expected to know what these are. They usually include:

- taking reasonable care of yourself and others in an engineering environment

- wearing appropriate clothing and using safety equipment as appropriate

- carrying out regular risk assessments

- following health and safety procedures and instructions

- keeping a safe, clean and tidy workplace

- ensuring that tools, equipment and machinery are properly maintained and fit for use.

Some of these, like keeping a workshop clean and tidy, are obvious, but others require a closer look.

Safety in the Workplace
Employee's Responsibility

Many people think that all the laws and regulations regarding safety at work apply only to the employer.

This is not the case: all employees have legal obligations at their place of work. There will also be requirements and expectations placed upon them by their employer. Sometimes these are included in the contract of employment.

The legal duties of an employee include:

- taking reasonable care for their own health and safety and that of others who may be affected by what they do or do not do

- co-operating with their employer on health and safety

- the correct use of work-related items provided by the employer, including personal protective equipment, in accordance with training or instructions

- not interfering with, or misusing, anything provided for their health and safety or welfare.

Employer's Responsibility

It is the responsibility of the employer to make sure that it is possible for an employee to fulfil the responsibilities above.

If employees are expected to wear specialist protective gear it is unreasonable to expect them to provide this themselves. An employer has a legal responsibility to protect the health, safety and welfare of all staff and is expected to take reasonable steps to do this. Ultimately, a safe and secure working environment will only exist if the employer and employee work together and co-operate.

As well as the HSWA, there are a number of other laws and regulations that the employer will be expected to follow. There will also be specific laws relating to the type of industry. This will range from hygiene regulations in food production to very strict rules relating to radioactive materials.

ACTIVITY

Write a full list of all the responsibilities that you have as a student in the workshops. In each case write a sentence explaining why.

Carrying Out Risk Assessments

A risk assessment is a way of looking at the working environment and assessing what things could cause harm to those working there. A safe working environment is not just important for the workers, it also shows that the company or organization takes its responsibilities seriously. It will never be possible to remove all risks from the workplace.

A risk assessment should identify aspects of the working environment that:

- might pose a real and immediate danger to staff. These need to be dealt with straight away

- pose a risk that is manageable as long as staff are properly informed and suitably trained

- pose little or no risk.

The risk assessment itself is useless unless it prompts immediate action. In many cases, risks will be identified and, although they cannot be eliminated, steps can be taken to reduce the likelihood of injury or damage.

Fig 2.70 Safety signs

How to assess the risks:

- look for the hazards

- decide who might be harmed and how

- under what circumstances could this happen?

- what precautions already exist?

- are they satisfactory, if not how could the risk be reduced?

- record your findings

- review your assessment and revise it if necessary.

If you feel further action is needed this should be clearly stated in your assessment. It may not be your responsibility to implement changes in the workplace but you have a responsibility to yourself and your colleagues to report any potential dangers.

When looking at hazards and risks you will need to consider everyone who comes into the workplace, however infrequently. They could include:

- young workers or trainees

- new and expectant mothers.

- those with medical conditions (for example, people with asthma or breathing problems may be affected by an environment with a lot of dust)

- cleaners, visitors, contractors.

It is very likely that the staff working in any hazardous environment will be informed and educated of relevant risks, but anyone who is new, or an infrequent visitor to the site, will not have that knowledge.

KEY WORD

Duties   tasks that are your responsibility

PORTFOLIO NOTES

Throughout this section of the book, there have been references to risk assessments that should be recorded in your portfolio. You need to demonstrate that you appreciate the importance of health and safety and that you understand how to manage it. The best way to do this is to assess the risks involved in every stage of your work and follow the correct procedures to minimize them.

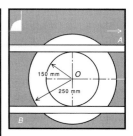

3 | Application of Technology

CHAPTER AIMS & INTRODUCTION

This chapter will explore:

- information and communications technology
- modern materials
- control technology.

The role of information and communications technology (ICT) has become more important than anyone could have predicted 50 years ago:

> There is no reason for any individual to have a computer in his home.
>
> *Ken Olsen (1926–), President, Digital Equipment, 1977*

Since the Second World War, the development of computers and computer technology has changed the world of work and society in general. The use of computers has meant that people can work faster and more reliably than they could before. Machinery can be controlled without workers. Mathematical problems can be worked out quickly. Drawings can be created and sent around the world in seconds. The computer is now commonplace and is used for numerous different tasks in the world of work.

The development of ICT has changed the role of the worker. Some companies have found that using ICT has enabled them to employ fewer workers. This can save money, as the company will be paying a smaller number of staff. However, ICT equipment and maintenance can be very expensive.

The aim of this chapter is to show how manufacturing and engineering has become more efficient. Modern approaches, processes, materials and techniques have been developed to help improve the production of goods.

ICT has been one of the major influences on modern manufacturing and engineering. ICT affects all aspects of production, from researching suppliers on the Internet, to producing working drawings on a computer screen, to controlling devices in a workshop, and in every other aspect of designing and making things. There are many books and websites that you can use to learn more about ICT.

Fig 3.1 Using technology

ACTIVITIES

Select a product that has been developed or changed in design over recent years.

1. Describe your chosen product. Use illustrations, dimensions and any advertising you can find.

2. Describe how your chosen product has changed over recent years. You could try to produce a timeline.

3. Describe the ways in which ICT may have affected how your chosen product has been developed.

REVISION NOTES

Computers are used to:

- design things

- make things

- control things

- communicate ideas between people.

REVISION QUESTIONS

1. Choose one way in which ICT is used in modern manufacturing or engineering. Explain how the job would have been done before computers were used. Why do you think computers are now used to do that job?

2. Computers now do jobs that would be dangerous for humans. Give an example of this.

① MANUFACTURING AND ENGINEERING SECTORS

Manufacturing sectors

- Biological and chemical
- Engineering fabrication
- Food and drink
- Paper and board
- Printing and publishing
- Textiles and clothing

Engineering sectors

- Aeronautical
- Automotive
- Civil
- Computer
- Construction
- Electrical and electronic
- Fluid

- Marine
- Mechanical
- Process control
- Telecommunications

You will need to find out what sort of items are designed and made by each of the manufacturing and engineering sectors listed above. Each sector produces a number of things. There are also lots of items that are made by combining objects from different sectors. For example, a car is made by the automotive sector. However, the carpets for the car are made in the textiles and clothing sector. The engine management system is made in the electrical and electronic sector. The braking system uses technology developed in the fluid sector and the handbook will be produced by companies in the printing and publishing sector.

Fig 3.2 Examples of objects produced by different industries

Other items may be developed within a single sector, but they may use equipment that has been developed in other sectors. For example a T-shirt will be made in the textiles and clothing sector, but the machines used to make the shirt will have been made in other sectors. The finished shirts will then be delivered to shops using vehicles made in the automotive sector.

It is the co-operative nature of manufacturing and engineering that enables successful products to be developed.

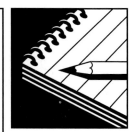

ACTIVITY

Describe the types of product manufactured by the manufacturing and engineering sectors. Give an example for each.

Sector	Description	Example

REVISION NOTES

You would not be expected to remember every product produced by every manufacturing sector industry, but you need to understand the types of thing each industry might make.

Control Technology

At the start of the Industrial Revolution, machines were invented that could carry out repetitive or dangerous jobs. In recent years the development of technology has allowed more complicated machines to be made.

Manufacturing machines can now work together to carry out a number of operations in a sequence, passing a piece of work from one specialist machine to another. Constant advances in control technology have made automated machinery possible, and many people now believe that the 'worker' is no longer needed. However, this is not the case as the majority of manufacturing and engineering is still carried out by skilled individuals aided by machines.

Control technology can be used to monitor pieces of equipment, such as drilling machines. Sensors check the machine and feed the data back into the processing system. Using this data, decisions can be made as to whether drill bits are operating at the right speeds and holes are being drilled accurately. The piece of work can also be checked using optical sensors, X-rays, and ultrasound; the results will once again be fed back to the processing system. If this process is linked to other machines in a production line, a fully automated system can be developed.

One area where this sort of control system has been used to benefit the workforce and the customer is the food processing industry. Food made at home is usually made by hand, however, when companies are manufacturing many millions of food items, e.g. biscuits, each item must be identical. In many food-processing factories there is a range of control systems in operation:

- refrigeration units – temperature monitoring

- ingredient transportation – conveyor systems, weight sensors and foreign body sensors

- cooking – temperature monitoring, time monitoring

- packaging – weight, volume and presence sensors used in combining goods with the correct packaging.

Fig 3.3 Control technology monitors this meat packing process

Each stage of manufacturing is monitored electronically and results are fed back to a central processing system. Workers oversee the process and confirm the results from the sensors.

ACTIVITY

Investigate the use of control technology in the production of one of the following:

- a portable stereo

- a mobile telephone

- a home entertainment device – DVD player, video recorder, TV

- a personal computer

- kitchen units

- clothing

- food products.

KEY WORDS

Computer system   input, output, processing and feedback

Sensor   a device that senses something, such as temperature or pressure, and feeds it into a control system as an input

Feedback   using data output from the computer system to change the input; for example, changing the speed of a drill because the sensor feeds back information that the work is getting too hot

INFORMATION AND COMMUNICATIONS TECHNOLOGY ②

Storing and Handling Information and Data

If you look at a string of numbers like this, 8, 6, 3, 9, 8, 7, it doesn't make any sense. We don't know what the numbers mean. They are data. If we are told that they are marks out of 10 for a product in a survey, we can look at them and decide that the product is quite popular. They are now in context, and have become information. We can use information to help us decide what to do.

Data manipulation has become essential in all industries. Products are only developed and made after extensive research to make sure that the customer will want the product. A specialist market research company could carry this out or, for a specialist product, the manufacturing company may do the research.

The data that is collected from the market research must be stored in a useful format. The best form of data for this task is a database or spreadsheet. Both of these formats allow the data to be organized, manipulated and viewed. These systems also allow the user to process the data to make information.

Fig 3.4 Data from market research is stored on an electronic database for easy access

A database is basically an electronic filing system. It is a collection of information organized in a way that makes it easier for a computer to select a particular piece of data. Databases are structured into fields, records and files:

- A field is a single piece of data – for instance, a name, or telephone number.

- A record is a collection of related fields – the name, address and telephone number of a customer.
- A file is the database itself, usually stored with a distinctive name – 'customers'.

Databases are good at sorting information and searching it. They might, tell us, for instance, that a certain product is more popular with men than women. Databases can produce reports based on the data they are storing, perhaps of all the customers grouped according to which area of the country they live in.

A spreadsheet is a large table of cells. Each of these cells can contain electronic data in the form of text, numbers, and formulas. A spreadsheet is a useful format as it can perform functions on the data within a cell, such as calculations. They can be used for invoices and to analyse numerical data. They could plot charts of quality data that would make it easier to see when a product run was starting to produce too many scrap components.

ACTIVITIES

1. Produce a database structure to record customer details.

2. Produce a report detailing how a company could use a spreadsheet application to help record stock details. Produce a form for recording information that could then be entered into a suitable spreadsheet.

KEY WORDS

Data   raw, unprocessed text or numbers

Information   data put into context so that it has meaning and we can make decisions based on it

Database   software that stores, searches and sorts data and information

Spreadsheet   software that performs calculations and plots graphs

Computer-aided Design Techniques

Computer-aided design (CAD) has increased in popularity as computers have developed, making them more powerful and faster at manipulating data. CAD software is quite demanding in terms of the hardware it needs to run on.

There is a wide variety of CAD software packages. Some of these packages can be used on standard computers. However, many of the packages used in industry require specialist computers, with a range of input and output devices. These industrial packages are often written especially for a particular company, and are used by specially-trained workers.

Fig 3.5 Some CAD packages can be used on standard computers

Whichever CAD package is used there are certain similarities:

- There is a drawing area.
- There is a library of commonly-used components.
- Lines and text can be formatted.
- Various views of the design can be seen and developed at the same time.
- Material volumes and weights can be calculated.

The difference is often in the output: more advanced systems will enable 2D images to be developed into 3D images with surface texture, colour and simulated materials. They will also allow the 3D representation to be moved or manipulated in 'real-time'.

There are many benefits to using high quality CAD to develop ideas:

- Rapid prototyping – enabling on-screen models to be manipulated in order to see what the final design will look like, or how it will behave in practice. This means that there is no need to make expensive models.

- Non-destructive and destructive testing – testing an object under load in order to study the object's tolerances or breaking point. Again this means that expensive models don't need to be built.

- Ease of communication – through the use of email and the Internet it is possible for designs to be distributed around the world almost instantly. This means that companies can use designers from across the globe to work together to develop the best possible solution. There is no need for costly trips around the world for meetings. The standards used to produce the designs have become international, as it is essential that anyone having to read a drawing can understand the symbols and layout.

- Ease of storage – A0 engineering drawings are extremely large and bulky to store, and a complex product might need dozens of them. Computers can store the drawings in a much smaller space, and only print them when they are needed.

CAD packages need high-resolution monitors to work well, and lots of processing power to make them work quickly. They often use special hardware, such as a graphics tablet rather than a mouse and a plotter that can print huge drawings rather than a standard printer.

ACTIVITIES

1. Produce a design using a CAD software package.

2. Produce a working drawing using drawing equipment.

3. Compare the results from the two methods of producing diagrams.

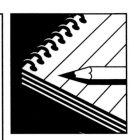

REVISION NOTES

- CAD means computer-aided design, and it involves a drawing package that runs on powerful computers.

- CAD packages can be very specialized and often use specialist input and output devices.

- CAD packages need skilled designers to work with them, but have a range of tools to make the designers' work easier.

Computer-aided Manufacture

As with computer-aided design, computer-aided manufacture (CAM) has developed as computing power has grown. However, many of the advances in CAM have come about through discrete control systems developed for particular devices. This has lead to machinery being developed that can be controlled by an 'on-board' computer.

For example, a CNC (computer numerically controlled) lathe has the basic elements of a standard lathe, with the addition of a computer that is used to control the spindle and tool movement. This control computer can be programmed directly. It is possible to design a product using CAD and then download it directly to the lathe, in a form the on-board computer can understand. The lathe will then carry out a sequence of operations, as defined by the design, to manufacture the component, changing feed rates and even tools automatically, when it needs to.

Fig 3.6 Using CAM on the factory floor

This method of working can be seen in many different industries:

- Robotic welding equipment in car manufacture plants.

- Computer-controlled biscuit cutters in food processing plants.

- Computer-controlled textile cutters in clothing manufacturing.

- Sheet material cutters in flat-pack kitchen manufacturing.

In many industries, a range of CAD systems is linked to a range of CAM systems to achieve an integrated manufacturing or engineering plant. In some companies this may be a number of linked plants worldwide.

CAM systems have changed the pattern of employment in some industries, because the computer takes over some of the machining decisions. It might mean that the company needs more people with computer skills and fewer people with machining skills.

The construction of a satellite and its launch rocket takes place in many different countries all over the world, with various different companies and institutions working together to produce components. The designs can then be sent electronically to the manufacturing or engineering plants, and are transferred to the CAM equipment to produce the parts.

ACTIVITIES

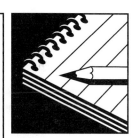

1. Carry out a site visit to a company that uses CAM equipment.

2. Produce a case study of approximately 500 words (include images), explaining how the company uses CAD.

N.B. If you are unable to make a site visit, there are a number of videos and web sites that can provide details that you could use.

REVISION NOTES

- CAM means computer-aided manufacture.

- CAD systems can be linked to CAM systems so that a product can be designed on computer and then manufactured by machines controlled by on-board computers.

- CAM tends to mean the company needs more computer specialists and fewer machinists. They still need skilled workers, but the skills they need have changed.

Communications Technology

Modern industry could not function without complex and reliable communications technology. The ability to communicate with staff in other parts of the building, off-site, or overseas has meant that production can be streamlined. This makes the process of designing and making something much more efficient.

Many companies rely heavily on telephone and email technology to communicate with their employees. This is very useful if they are working together on a project in different locations. Text messages are also useful to send short messages to workers on the move.

The ability to talk to someone in another building is now taken for granted in most modern companies. Video conferencing is also starting to make a difference to how workers communicate. Meetings can be held with people from all over the world and nobody has to leave their desk. If pictures are not needed then instant messenger services allow people to type messages in turn and hold a text-based 'conversation' in real time.

Fig 3.7 Video conferencing technology enables long distance, face to face communication

Video conferencing technology can also be used to monitor a production process. A video camera can be set up to record a machine in operation and play the video in real-time to a computer or TV screen in another part of the company. Staff can view the video, which can be very useful for training employees. It can also allow an operator to monitor more than one piece of equipment. One great benefit is that it can allow an operator to watch a process taking place from a position they would not be able to see normally, such as from inside a machine or in a hazardous environment.

Manufacturing and engineering companies also need to transport their goods and keep in contact with service engineers. Both of these operations have been greatly altered by the mobile telephone. Deliveries of goods and services can now be more efficiently coordinated with the increased availability of satellite navigation systems.

It is also possible to monitor goods after they have been passed to the customer, or once the goods are in use. Formula 1 racing cars have radio links that send data from sensors all over the car back to the pit crew. The technicians can then make adjustments to the fuel or brakes without having to stop the car. Similar systems are now starting to appear in some road cars. One prestige company monitors its top of the range cars after they have been sold, and can send instructions to the engine management system anywhere in the world. It can also use the system to track a car if it has been stolen.

REVISION NOTES

Through advances in technology, employees can communicate using:

- telephone (land line or mobile)

- email (they can also attach files)

- text messages

- instant message services (chat)

- video conferencing.

REVISION QUESTIONS

1. Describe two effects that improvements in the telecommunications sector have had on manufacturing or engineering.

2. Explain why a service engineer who works on a number of sites, may feel that the mobile telephone has not made his job easier, but his manager may think the opposite.

?

Computer Technology

Engineers and manufacturers have found that the advent of the computer has enabled them to work faster and more efficiently. This is due to developments in microprocessor and memory devices.

Computers first started to be used commercially just after the Second World War, but they were incredibly expensive and large. In the 1970s the personal computer or PC was developed.

Fig 3.8 The personal computer was developed in the 1970s

This came about through the use of microprocessors. These are small devices that carry out many calculations per second. Previous to this computers had used big transistors as switches. This meant that

they were large and used a lot of electricity. They also produced a great deal of heat, so they had to be kept in air-conditioned rooms.

As microprocessor speeds have increased, so has the capacity to store information. It is generally said that computers move on a generation every 18 months or so. This means that a brand new computer, bought today, will be out of date in 18 months! A generation, in computer terms, is a doubling of processor speed. Many PCs can now run at 3 GHz – 3 million calculations per second. In 18 months time we may start to see 6 GHz machines. However, much of what people use PCs for could be carried out on a machine running at around 100 Hz – 100 calculations per second.

Memory devices have developed in a similar way. Although memory size can mean lots of different things, depending on what it is used for, the speed of accessing stored data has increased in line with the speed of microprocessors. This means that, for a set amount of money, a computer is now faster and can hold and retrieve more information than a computer costing the same amount in the past.

This is the same for larger industrial systems. A company may spend millions of pounds on a new computer system, which in 18 months will be out of date. However, it is unlikely to be able to invest a similar amount to keep up to date. This means that the system will probably be upgraded rather than replaced.

In engineering and manufacturing this can lead to problems. As new technology becomes available, a company has to make decisions as to whether to invest in the new technology, or stick with what it has already. Good planning is essential, and it is wise to 'future proof' when you buy a computer, by choosing one that does more than you think you need.

ACTIVITIES

1. Produce a timeline charting the increase in processing power of computers – there are many good books and web sites that can provide the information you need.

2. Investigate the range of memory devices available that could be used to store and transport computer data.

3. Produce a multimedia report of your findings.

REVISION NOTES

- The 'brain' of a computer is its processor or CPU.

- Computers are constantly getting faster and can become out of date in 18 months.

- As new computers are built, they can store more data and access it more quickly.

- Replacing computers is expensive and companies will usually try to upgrade them first.

Micro-Electronic Components

Many engineered and manufactured products incorporate micro-electronics. In the 1960s, the race to put man on the moon led to the miniaturization of a number of electronic components – the integrated circuit (IC). This in turn led to the combining of these components into devices that could do more than one thing, for example, the portable cassette player or Walkman.

In the 1960s, recorded sound could be stored on tape, but not on cassettes. The tape used was a reel-to-reel system, which involved two reels, approximately 10 cm in diameter, being played on a machine that was about the size of a laptop computer. As any space travel is concerned with moving weight from earth into space, the idea of taking large and heavy equipment was a problem. The compact cassette we now use was developed alongside a small, lightweight player. In the years following the space missions, this device has become smaller, lighter and able to carry out more functions, but basically it is still the same thing. The control circuitry has been combined into one IC. This has been made smaller than the earlier versions and it now uses even less power. The integration of circuitry into one device is what brought about the personal cassette player and many other similar objects.

Fig 3.9 Space travel led to the development of small, lightweight electronic equipment

As technology has continued to move on, there are now personal stereo systems with hard discs. These can store many hours worth of music, or any other digital data, such as video. With this capacity, some companies have developed display devices that can allow the user to view video files stored on the hard discs. These devices are designed for recreation, but are similar to devices used in industry to transport large data files securely.

The developments in display devices have meant that many machines now use a screen to display control features, or to show data from monitoring sensors. Much more data can be shown clearly using a screen rather than through a number of red and green lights, as were used in the past.

These developments have come about through the use of new materials and processes, as well as the creativity of the designers and inventors.

ACTIVITIES

Many personal electronic devices have become more compact over recent years. Mobile telephones have undergone a transformation similar to that of the personal stereo.

1. Carry out an investigation into the developments that have happened to mobile telephones over the last 15 years.

2. Explain how the miniaturization of the components has made using the mobile phone more 'user friendly'.

REVISION NOTES

- The development of microchips, or integrated circuits, has meant that electronic devices are a lot smaller than they used to be.

- Many technologies were developed for use in the space programme, but are now being used in everyday life.

③ NEW COMPONENTS AND MODERN MATERIALS

Modern manufacturing and engineering has developed new ways to design and make things. Designers also have access to materials and components that did not exist in the past.

Throughout history we have lived through 'ages':

*   The Stone Age – when man worked with stone implements and used stone to build with.

*   The Bronze Age – when man developed metallic bronze to make bronze tools and implements.

*   The Iron Age – when man made things from iron-based metals.

*   Today we are living in the Technological Age – we can use technology to help us improve our lives.

Fig 3.10 Stone age man used implements made from stone

Technology has enabled the development of new materials such as plastics, polymers and composite materials. Composite materials use the properties of more than one substance to produce a more useful material, such as:

*   Carbon fibre – uses the strength and weightlessness of carbon, which is formed into plastic resins and makes the material very durable.

*   Metallic alloys – use the properties of different metals, which are combined together to produce very strong, yet lightweight materials that can be formed into complicated shapes.

*   Plastic sheet, plastizote, polypropylene, corriflute, etc. – the basis of these materials has been around for many years, but now they can be formed into large, brightly coloured sheets that are more durable.

These materials can be used to form new objects. To make these objects work, new components are needed. The components can be used to achieve more reliable and creative solutions to problems:

*   High power, lightweight electronic power cells.

*   Lighting components, such as LEDs, halogen bulbs, and energy saving bulbs.

*   Powerful integrated circuits with high capacity memory chips.

*   Corrosion resistant valve components for hydraulic and pneumatic equipment.

Smart materials are being developed to allow electronic circuits to be built into fibres, so that in the future a mountaineer's hat might be able to tell him when he is getting dangerously cold, because the fibres of the hat contained a heat sensor.

ACTIVITIES

A modern family car uses a range of different materials to produce strong and lightweight body parts. Many of these materials are difficult to investigate as they are often painted.

1. Carry out a research investigation into the materials used to make the bumpers of modern cars.

2. Describe how and why car bumpers have changed since the 1960s.

REVISION NOTES

* New technologies have produced new materials.

* New materials are often made by combining existing ones.

* New production methods can make old materials behave in different ways by changing their structure, e.g. carbon fibre.

REVISION QUESTIONS

1. What is an alloy?
2. Give one example of how an alloy can be more useful than a pure metal.

?

Polymers, Plastics, Adhesives and Coatings

What is a 'polymer'? The word polymer comes from the Greek 'poly' meaning many, and 'meros' meaning parts or units. A polymer is a chain-like molecule made up of smaller units, 'monomers'. The monomers, which are made up of atoms, bond together. This bonding makes a chain; many chains are contained in a section of material. The length and pattern of these chains and the chemicals used to produce them affect the properties of the material.

Polymers include many types of industrial materials. Polymer is often used as a term for plastic, but many biological and inorganic molecules are also polymeric. All plastics are polymers, but not all polymers are plastics.

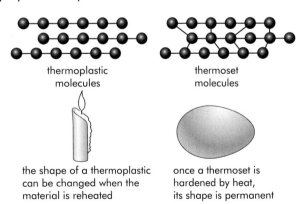

thermoplastic molecules thermoset molecules

the shape of a thermoplastic can be changed when the material is reheated

once a thermoset is hardened by heat, its shape is permanent

Fig 3.11 The chemical linking of polymers

Plastic refers to the way a material behaves under force or when it melts. Commercial polymers are formed through chemical reactions in large vessels under heat and pressure. Other ingredients are added to control how the polymer is formed and to produce the proper molecular length and desired properties. This chemical process is called 'polymerization'.

There are many different types of polymer material, and they have each been developed to do different things. Some polymers, such as acrylic, are very strong in sheet form. Acrylic is also very clear, clearer than glass. Sealife centres use acrylic for the walls of large fish tanks, as they need to be very thick to withstand the pressure from millions of gallons of water. Some of the tank walls can be up to 33 cm thick and still transparent. Glass would become opaque at that thickness: you would not be able to see through it.

Other polymers can be used as adhesives as they are in liquid form at room temperature, yet when mixed with a particular catalyst they 'go off', making them hard. This property is useful in adhesives as the liquid can penetrate the surface of two materials, then when it sets it can hold the two materials together. This property is also used for coating materials, such as paint, as the liquid state bonds to the surface of the material, and then as it dries it goes hard.

ACTIVITIES

1. Make a list of five polymer materials that you could collect examples of.

2. Collect samples of each.

3. Carry out three tests on each sample to investigate:

   a. hardness

   b. transparency

   c. conductivity – heat and electricity.

4. Write up your findings as a report or on-screen presentation.

REVISION NOTES

- Polymers are long chain-like molecules.

- All plastics are polymers, but polymers can also be other things.

- Polymers can be used as solid materials like acrylic, or as glues and paints.

Metals and Composites

Many metals are mined; they are dug up out of the ground and then refined to create a pure metal. However pure metals are very rare, and are often not very useful. To make metals more useful, their properties are altered by combining them with other metals and chemicals. The resultant material is then called an alloy. Mild steel is an alloy of iron and carbon, but other metals such as chromium, molybdenum and nickel can be used to improve its properties.

Fig 3.12 Many metals are mined – these are pure metals

By mixing materials together, the properties of the materials are combined. For instance iron, which rusts in damp environments, can be made resistant to corrosion. It can also be made harder wearing and lighter in colour. Many alloys can be recycled and the separate materials reused. A similar effect can be found in using plastic materials. Combining properties will create a substance that has the benefits of the materials it is made from.

Glass-reinforced plastic (GRP) is commonly used for small boat hulls. It is made from a liquid gel plastic with strands of glass embedded in it. The plastic resin is waterproof and easily formed, but not very strong, in fact it tends to be very brittle. The glass is used to add strength, as the strands create a mesh that holds the resin together. The two materials cannot be separated after manufacture, as they have become a single material, a composite material.

Modern developments in materials technology have lead to the introduction of materials that bridge the line between plastics and metals. These often involve mixing metals with ceramics or polymers, to achieve a combination of properties from the two materials.

Terfonol is a material that can be made to change shape using a magnetic field. As a magnetic field is applied to the material it can get bigger or smaller. This is dependent upon the polarity of the magnet and the strength of the magnetic field. This technique can be used to make fine adjustments to machinery extremely quickly.

NiTinol materials are metals that behave like thermoplastics. They can be formed and hold their shape, then heated and reformed, without damaging the material composition.

ACTIVITY

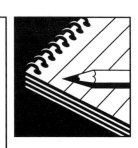

Produce a table of metals, filling in the answers to the following questions.

a. Is the metal ferrous – iron-based?

b. Is it an alloy – a mixture of more than one metal?

c. Does it conduct electricity?

d. Can it be magnetized?

Include notes on anything else you may have found out.

Metal name	Ferrous?	Alloy?	Conductor?	Magnet?	Notes

REVISION NOTES

- Metals can be mixed together to form alloys.

- Alloys are made to improve the properties of pure metals.

- Non-metals can also be combined, and metals can be combined with non-metals. If the materials are mixed, they can be separated and reused, but if they form a composite, they cannot.

Modified Food Ingredients and Methods of Preparation

Throughout history people have tried to develop ways of feeding themselves more efficiently. Modern technology has enabled scientists to find ways of helping food producers to improve their yields. Crop and animal farming has been heavily affected by a number of modern scientific developments. The biological and chemical manufacturing sectors have become increasingly important to the production of food.

Since the discovery of genetics and DNA, experts have been trying to improve the abilities of crops to defend themselves against disease, produce more fruit, or grow more quickly. This has been achieved in a similar way to the development of new alloys and composite materials. Seeds from different plants have their DNA altered by the introduction of other genetic material or DNA from plants that have different properties. The resultant seed then contains the characteristics of both plants.

Imagine one plant produces small fruit and is resistant to disease and another produces large fruit, but is not resistant to disease. The 'hybrid' will have characteristics from both: it will produce larger fruit and be more resistant to disease.

Fig 3.13 A genetically modified crop

Animal feed has been studied by scientists to enable farmers to produce more meat per animal. This does not mean just bigger animals, but growth in particular areas to increase muscle density.

There are also a number of developments in the way food products are treated after they have been grown. Radiation can be used to kill bacteria and make fresh produce last longer. Packaging materials can be used that allow fresh produce to continue to ripen, therefore enabling produce to be picked earlier.

Once the farm has produced the raw materials (crops, fruit or meat), the conversion into food can involve engineers and manufacturers.

Much of the process of food production is now automated, or is made on a production line. The system of producing steak and kidney pies rarely involves any manual work. Raw materials arrive at the factory. The flour and fat are combined to produce the pastry. The meat and other ingredients are chopped and cooked. Then the pastry is cut and inserted into trays, where the meat filling is poured in. The top is then added and the whole pie is then passed through an oven. All of these stages are automated. The machinery is monitored by control technology so alterations to ingredients or temperatures can be carried out remotely. The finished product is then packaged ready for distribution without ever being touched by human hands.

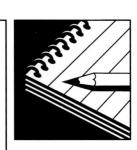

ACTIVITIES

1. Produce a table showing the advantages and disadvantages of genetically modifying foodstuffs.

2. Investigate and produce a flow diagram for the production of a food item on an industrial scale. You may like to consider:

   - biscuits

   - soft drinks

   - bread

   - sandwiches

   - TV dinners (microwaveable ready meals).

REVISION NOTES

- Scientists and engineers are involved with food production both at the growing stage and the manufacturing stage.

- Genetically-modified crops may have better properties than the originals, but many customers are not happy about eating them.

- Animals can be bred or treated with chemicals for more economical meat.

- Organic farmers avoid all these methods.

Textile Technology

Textiles have been used throughout history. Many industrial advances have developed through the textiles industry, and then been adapted to be used in other industries. The automated loom that was designed by Jacquard using punched cards to 'program' the loom became the basis for the method of programming computers. Factory designs for the production of textiles were later adapted by many industries as 'production lines'. With modern advances in materials, textile technology has again become important to many other fields.

The development of liquid crystal displays (LCDs) has meant that, via an electrical current, images can be shown on yarns and fabrics. Until recently any display had to be mounted onto a rigid material, usually glass. This meant that displays were quite large and heavy. The development of textile technology to be able to incorporate LCDs has meant that displays can be on flexible, lightweight materials. This means that designers can build displays into different objects, such as clothing. There are a number of companies developing 'wearable' computers. The drives and processor units are discreetly sewn into the garments and the display is attached to the sleeve. There are also a number of mobile phone designs being developed with screens that are curved, or incorporated into objects such as clothing or bags.

Fig 3.14 Companies that make trainers will use textile technology

Other uses for this type of textile technology also include photo-sensitive panels that can be used as blinds for windows. As the sunlight increases on the blind, an electrical current can be passed through the material, turning it more opaque thereby stopping the sunlight passing through.

Another new concept in textile design is the use of dyes that change colour when the temperature changes. A common use of this is in thermometers for use with small children. The thermometer is a strip of material containing an area that has a thermochromic dye applied to it. At certain temperatures it is one colour, but at a few degrees above or below it changes colour. This can give an indication of body or bath water temperature, without the use of traditional glass thermometers. These thermochromic dyes can also be used in clothes, for fashion or in safety equipment where the user needs to know what the temperature is.

ACTIVITIES

1. Produce a number of illustrations showing how clothing could look with LCDs and thermochromic dyes.

2. Describe how these technologies could improve safety for workers.

— REVISION NOTES —

- Microprocessors and displays are being developed that can be built into fabrics.

- They can be used for fashion or for practical uses. A mobile phone in your hat might be fun for you but save a soldier's life.

- Thermochromic dyes can also be used for fun or for safety. They change colour depending on the temperature, so could be used to show if a person's body temperature is dangerously low.

SYSTEMS AND CONTROL TECHNOLOGY ④

Process/Quality Control and Automation

Many industrial devices now have computers built into them. These devices carry out a number of functions without an operator making adjustments. Many of these embedded computers use programmable logic controllers (PLCs). These are integrated circuits that can be programmed to carry out certain routines. One type of PLC can be used for many different applications.

The control of a dishwasher, for example, is similar to the control needed for a washing machine. The embedded computer has to make components within the machine do things at certain points in the programme. This might happen at specific times or when sensors give certain readings.

The PLC is programmed to carry out a set sequence of actions. The PLC must start a timer running, open or close particular valves, monitor temperature, turn a heater on and off and so on. The PLC is then inserted into the machine's circuits. Because the PLC in a number of devices is similar, the cost of components is kept lower. As the other components in the machines need to be able to communicate with the PLC, they are also becoming more uniform, again reducing costs to the manufacturer.

PLCs are becoming more widely used in machines and other objects. Many industrial pieces of equipment require simple monitoring and control, which can be carried out by an embedded computer using a PLC. A number of PLC devices can be connected to a master computer which can make adjustments to them if it needs to. This is a great advantage in an automated production line, where processes are repetitive. Each machine can be monitored and adjusted to ensure that the production line continues to work at an optimum level. This helps to avoid problems occurring that would stop production or cause a drop in the quality of the goods being produced.

ACTIVITIES

1. Describe how important embedded computers or PLCs would be to a robotic system, such as the Rover that is carrying out tests on the surface of the planet Mars.

2. If your school and college has equipment capable of programming PLCs, design and make a robotic device that uses a PLC.

REVISION NOTES

- Programmable logic controllers, or PLCs, can be built into many machines.

- They control the operation of the components of the machine, turning things on and off when they need to.

- They become cheaper if they can be used in more than one machine, because parts can be standardized and do not cost as much to make.

- Outputs from several PLCs can be fed into a master computer that can make adjustments when it needs to.

Robotics

The term 'robot' was first used in a play by Czech playwright Karel Capek in 1921. He used the term to describe artificial workers, not machines. At the time that he wrote his play, there was still very little automation in factories. Since then factories have become increasingly reliant on technology for repetitive tasks because of the development of control systems.

Fig 3.15 A robot from the 1960s

Modern automated production methods require that machines carry out tasks in set sequences over long periods of time. This is called continuous operation and it can create complex problems. A lathe working continuously for many hours will wear down its cutting tool. It is important that the tool can be monitored and adjusted or changed if necessary. Automatic tool changes without stopping the machine may be possible.

Robotic systems that move, such as welding equipment, need to return to a precise position after each movement. Again this will need close monitoring and adjustments to ensure that the machine movement stays within tolerances. This continued monitoring and fine-tuning of machinery makes it more likely that the items produced will be identical in size and shape. This is called 'reproducibility'. If large numbers of identical components are needed, a highly monitored system should be used. The use of such systems can decrease the time required to produce a batch of items, as the production line can work for longer periods without having to stop for maintenance or adjustments to be made by workers.

Using robotic systems instead of people can also be a great advantage when operations need to be carried out in hazardous environments where extremes of temperature or fumes and gases could be harmful to workers.

ACTIVITIES

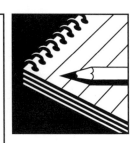

1. Carry out an investigation and produce a report into a local industry that uses automated or robotic production methods.

2. Produce a report about the differences you can find between a company that uses robotic manufacturing systems and one that does not.

REVISION NOTES

- Robotic systems are good at repetitive tasks carried out over a long period of time.

- Continuous operation means producing goods without stopping the production lines.

- Robots can work in conditions that would be dangerous for humans.

- Robotic arms can carry out tasks very precisely but need regular checks to make sure they are calibrated properly.

ICT As Applied to Integrated Manufacturing and Engineering Systems

As mentioned in earlier sections ICT is an essential part of modern manufacturing and engineering. A product is developed using ICT as a tool to aid design, production, transport, distribution and sales.

Computer Integrated Engineering (CIE) combines the use of CAD to design products and CAM to control the machines that make them. Computer Integrated Manufacturing (CIM) covers the same areas, although the main difference between the two systems is that manufacturing is involved with making large numbers of products. This means that the equipment used is much more likely to be concerned with assembling components to produce a product, rather than dealing with raw materials.

Fig 3.16 Robotic systems in use

There are many similarities in the way ICT is used in engineering and manufacturing:

- Design – the use of CAD, communicating designs to others.

- Marketing – market research, manipulation of results from research.

- Production planning – flow charts, Gantt charts and flow analysis.

- Material supply and control – Internet sites for suppliers, monitoring of stock, Just in Time systems.

- Processing and production – monitoring of equipment and automation.

- Assembly and finishing – robotic assembly plants and the application of new materials.

- Packaging and dispatch – stock control, telematics to control distribution vehicles.

ACTIVITY

Complete the table opposite to describe how ICT is used at each point of the production cycle.

Stage	Example of ICT	How	Example of ICT	How
Design	The use of CAD		Communicating designs to others	
Marketing	Market research		Manipulation of results from research	
Production planning	Flow charts		Gantt charts	
Material supply and control	Internet sites for suppliers		Monitoring of stock	
Processing/production	Monitoring of equipment		Automation	
Assembly and finishing	Robotic assembly plants		Application of new materials	
Packaging and dispatch	Stock control		Telematics to control distribution vehicles	

REVISION NOTES

The table above should remind you of how ICT can be used in every part of production.

5 ▶ IMPACT OF MODERN TECHNOLOGY

Range, Types and Availability of Products

Modern engineering and manufacturing methods have enabled companies to produce an incredibly diverse range of goods, from foodstuffs using genetically-modified plants, to mobile telephones capable of sending and receiving live video messages.

Every sector of the engineering and manufacturing industry has developed new processes to make products. Many of these products are only made possible through the invention of new processes and materials. As technology has moved on, so has industry. In our society we are surrounded by goods and services that have been provided by the developments in engineering and manufacturing.

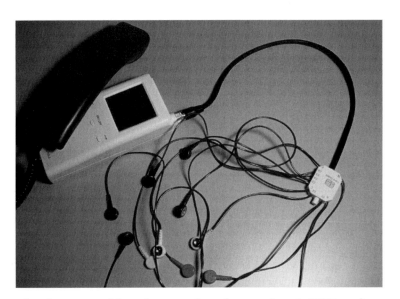

Fig 3.17 Remote technology enables the results of a patient's ECG to be sent to the hospital via a wireless connection

- Medical equipment that monitors an individual's health, remotely, via a wireless network connection to a hospital computer.

- Fruit which grows thousands of miles away, delivered to the supermarket in packaging that enables it to continue ripening whilst being stored.

- Cars that have on-board computer systems, monitoring the engine's performance and adjusting fuel use and lubrication.

- Personal entertainment devices that allow the user to watch full-length films, in colour, with stereo sound, while travelling.

- Mobile telephones that can be used anywhere in the world, even on top of Mount Everest.

- Jewellery that can change shape and colour depending on the mood of the wearer.

- Home computer systems that can be used to access the Internet, show movies or play games, as well as carry out complex business operations.

All of these use new materials or the application of new technology in order to satisfy the needs and desires of the customer.

Many of these products are commonplace now, but they are only available for as long as the technology stays available. Many of them require the use of complex new materials or processes that are themselves reliant upon other new technologies. Each sector of the engineering and manufacturing industry relies heavily on other sectors to produce successful and useful products. A shortage of raw materials or skills in one area can have a dramatic effect on other areas.

ACTIVITY

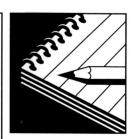

Choose a product from the list below. Produce a graphical representation of the dependencies that product has with other products and manufacturing sectors.

For example, a piece of jewellery is made using silver; the silver has to be mined and purified; the ring is then formed using machines made by another engineering sector; the finished ring is transported to the shop by road; the lorry was developed in the automotive sector, etc.

- Medical equipment
- Fruit
- Car
- Personal entertainment device
- Mobile telephone
- Item of jewellery
- Home computer system

REVISION NOTES

- Technology developments mean that new products are constantly appearing.
- Developments in one sector will have implications for other sectors; for instance, developments in biochemistry may help produce packaging that keeps food fresh longer, which helps the food industry.

Design and Development of Products

In the early part of the twentieth century a number of industrial designers got together and set up a design school. Until then most things were not designed, as we think of design now. Many of the items people used at home or at work were made to do a specific task. Products were often made as one-offs, or small batches, by hand or with machines controlled by workers.

As the century passed, a number of important things happened, including two world wars. Many great developments in technology come about through inventions first used in warfare: rockets, jet engines, portable radios, wristwatches, medical advances in pain relief, and many more. Materials were developed combining properties from a number of raw materials.

In the 1950s and 1960s, society rebelled against the dullness of the years immediately after the Second World War, when food and other items had been rationed by the government. This led society to a more optimistic view of the future of technology. USA and Russia embarked on a space exploration programme, both nations competing to get a man in space or on the moon.

Fig 3.18 This picture from the 1960s shows how everyday products, such as cars, have changed

In the 1970s personal computers were made possible, and most homes in the UK had cars, telephones and colour TV. Only a few years earlier these products had been considered luxuries. By the 1980s mobile telephones had appeared and computer systems were becoming more affordable and reliable. Industry started to become more reliant upon embedded computers and robotic control systems.

The new millennium has seen advances in food technology, genetic engineering of crops and the cloning of animals. Now medical engineering of replacement body parts is becoming more likely and cheap flights around the world are commonplace.

Many of the products we use now have come about through the developments of the last 100 years. As new materials and processes have become available, they have enabled engineers and manufacturers to produce more variety of products, satisfying the needs of the users.

ACTIVITIES

1. Make a site visit to a local engineering or manufacturing company that has been on the same premises for a number of years (over 15).

2. Carry out an interview with a member of staff who is able to describe how the facilities have changed over the years – this may be a member of staff who has worked there for many years or somebody with access to records.

3. Try to get some images of 'then' and 'now'.

4. Write up your interview in the form of a news report.

REVISION NOTES

- Over the course of the last century more people were able to buy products because they had more money.

- This meant that a design industry developed to make things which people wanted rather than needed.

- More products were made in factories, rather than by craftsmen working by hand.

- Many of these products were developed from technology originally used for something else – non-stick frying pans were developed from technology that was originally for the space programme.

- Some of the products produced are useful, some are fun and some may not prove to be popular with consumers.

Materials, Components and Ingredients

The introduction of new materials, components and ingredients has enabled designers to develop many new products, which has in turn lead to changes in production methods.

The advent of modern polymer technology has enabled designers to use more complex shapes in their designs. This has led engineering and manufacturing processes to be developed to make these designs. Injection moulding is the most common method for producing plastic items. The process has been used for many years, but with the introduction of new materials, the shape of the moulding can be made much more complex, and the product can be made stronger and lighter.

Components such as PLCs, or integrated circuits, have helped miniaturize products. Along with advances in materials, this has led to many devices becoming portable, for example, the personal stereo and laptop computer. A few years ago, such items would have been too large to be carried about.

Textile technology has led to the development of breathable textiles, which allow the body to lose excess heat while keeping the wearer dry. Armoured textiles protect police and military personnel all over the world. The advances in textile technology have made the materials stronger, yet lighter and more comfortable. Motorcyclists are protected by natural materials such as leather, but also extremely strong but flexible woven materials, such as carbon fibre.

Food is now much more varied than a few years ago and we can now obtain fresh ingredients from all over the world. This food can then be sold to the public or processed on an industrial level to produce high quality pre-prepared meals. The diet of an individual today can be much more varied than it would have been 20 years ago. Advances in food technology have led to a greater awareness of the body's needs. All food goods, apart from fresh produce must have a label telling users what ingredients have been used and the nutritional values of the food. This allows users to make more educated decisions over what they eat.

ACTIVITIES

1. Choose two products that you have studied and, for each product, describe how it has been affected by the use of:

   • information and communication technology

   • modern materials

   • modern production methods.

Product One:

ICT	
Modern materials	
Modern production methods	

Product Two:

ICT	
Modern materials	
Modern production methods	

2. Summarize your research in the form of index cards so that you will remember the main points for your exam.

Safety and Efficiency of Modern Methods of Production

As modern materials and processes have been developed, the efficiency of production has often increased. Automation of a production process, if carefully monitored, can improve efficiency in a number of ways:

- Less time is needed to produce a number of items, as the machines can work continuously, without lunch breaks or holidays.

- The accuracy of production can be improved, as the machine can be constantly monitored and the results from the monitoring used to make adjustments, without having to halt production.

Another advantage of modern methods of production is a more efficient use of materials:

- Computer software can be used to work out the best use for a sheet of material, working out where cuts need to be made to ensure minimum waste.

- Components can be accurately mounted onto a circuit board with robotic soldering systems.

- Stock can be monitored to ensure that anything likely to perish can be used before it becomes unusable.

- Storage conditions can be kept at an optimum state. Temperature, lighting and humidity can all be tracked by sensors.

Production costs can also be reduced through the careful management of energy consumption:

- Equipment can be monitored to ensure that it is running efficiently. If it is found to be using too much fuel, adjustments can be made to the operation or the machine can request a maintenance check.

- Equipment can be automatically shut down when not in use, or put in a standby mode, ready to be restarted when needed.

- Lubrication can be monitored, to ensure that any moving part does not use more energy than it needs.

Efficient transportation and distribution systems can also have a huge impact on the efficiency of production. There are very few empty lorries on the roads nowadays: haulage companies try to get a lorry to deliver its load and immediately pick up another load for distribution elsewhere, so that no time is wasted travelling without a load.

ACTIVITIES

1. Using information from a site visit, describe how efficient the production method seems to be. Try to spot where materials are transported for no apparent reason or production lines stop working because a single machine develops a problem.

2. Cars are often transported from one location to another on transporter wagons.

   - A transporter carries four cars.

   - Each car has an average fuel consumption of 40 miles per gallon.

   - The transporter has a fuel consumption of eight miles per gallon.

   a. Is it more efficient, in terms of fuel consumption, to transport the cars or to drive them to their new location?

   b. What could be the other factors that influence the use of the transporter, rather than driving each car?

REVISION NOTES

- Manufacturers and engineers are being encouraged to find more efficient ways of producing things.

- More efficient could mean using less energy or a smaller amount of raw material or a raw material that is easier to replace.

- Machines can be fitted with sensors that show when they are running too hot – a sure sign of wasted energy. This might mean they need oiling to improve lubrication.

Improved Characteristics of Products

The reliability and ease of use of many products has greatly improved due to the advent of new materials and production methods. Much more thought is put into the design of a product so that it will appeal to the customer, because there are often other products available from competitors that perform the same function.

There are hundreds of models of mobile phone, for example, each offering the same basic facility – the ability to make and receive a phone call when on the move. Each of the companies that make mobile phones has to try to make their product more useful or attractive than their competitor's version.

By using new technological developments to add extra functions, they may extend their appeal to different types of customers.

- They could make the phone smaller or lighter through the use of different materials.

- They could offer different finishes, again through the use of different materials.

- They could make the phone easier to use. This may cause a conflict with other aspects, such as size, but through using more on-screen or voice-activated technology, companies could overcome this.

Many new products also try to appeal to the environmentalist market. There are various laws that cover the product life cycle, some of which call for any product to be recycled in some way. Many car manufacturers now offer to recycle the vehicle at the end of its life, claiming that every component will be either recycled or reused.

Many products manufactured from polymers carry the recycle symbol and the type of plastic used. There are also instructions on most goods regarding the method of disposal that should be used.

A modern glass manufacturing plant in the UK recently moved to using 100% recycled glass. Many products are available that use recycled materials, but we still have a long way to go to achieve targets set by the government and EU about how much we should recycle.

There are many products that can now be passed to specialist companies who will reclaim useful resources from the spent product. For example power cells, the 'batteries' used in mobile phones, contain various metals that can be reclaimed and then reused.

ACTIVITY

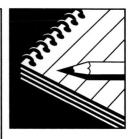

For one of the products you have studied, produce a report on how the design could be improved further considering the following:

- size

- weight/density

- ease of use

- disposal.

The report should include sketches of changes you would make.

REVISION NOTES

- Pressures on manufacturers have changed. 100 years ago, they would not have worried about how to dispose of their product, now they have to. New laws on safety, disposal of toxic waste, etc., have to be taken into account by designers of new products.

- Consumers are always looking for new things, so products have to be developed to keep up, or customers will buy what they want from other companies instead.

Markets for the Products

Before a product goes into production, a company must ensure that there is a potential customer. This is often done through market research. Many products made by engineering and manufacturing companies are sold to the public, following a 'perceived need'. This means that somebody felt that a particular product would be worth making and selling as they thought the public needed it, whether they actually did or not!

Many homes in the UK have gadgets and other items that, once bought and used on a couple of occasions, are put to one side and never used again. Was there really a need for that item? We often convince ourselves that we need something when, in fact, if it were not available we wouldn't have thought about needing it.

However, there are many other products that are produced because they do improve our lives in some way. Modern improvements in medical care, such as baby monitoring devices and so on, have made a dramatic difference to the life expectancy of an individual in the UK. Modern safety equipment has saved the lives of many people. Convenience foods have enabled people to produce meals quickly, using modern microwaves and pre-packed meals.

Fig 3.19 There is a large market for microwave meals in the UK

As technology moves on, the markets for selling goods can also change. Anyone buying a television today will probably not be interested in a black and white model, they will want the latest flat-screen, with surround sound and all the other possible extras.

This means that companies are always looking for new places to sell their goods. As countries around the world become more technological, each society's needs, or perceived needs, are met by engineering and manufacturing companies.

Due to the development of new materials and processes of manufacture, it is often relatively simple to alter a design so that it appeals to 'new' customers: Changing the words on a label so that it is in a different language, or changing the instructions on how to use a product can be done quickly through the use of ICT.

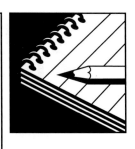

ACTIVITY

Gather information on three different consumer items from the list below:

- a mobile phone

- a low-calorie ready meal

- a piece of safety clothing or personal equipment

- a prescription drug

- a DVD

- a piece of sports equipment

- a personal entertainment device.

For each of the items describe in detail:

- which market it has been designed for

- what specific details have been incorporated to satisfy the needs of that market

- what adaptations could be incorporated to make the item appeal to other markets.

REVISION NOTES

- If a product is to be sold there has to be a market for it.

- Markets change as consumers demand new things.

- Market research can show what consumers are looking for.

- Sometimes a designer has an idea that proves really popular, even though it didn't show up in research. This might be because the public didn't know they wanted it.

- Other times, a product that seems like a good idea doesn't sell because people just don't use it as much as they expected to. A kitchen gadget that looked useful might prove to be really difficult to clean, for instance. It might sell well at first, but not for very long.

6 CHANGES IN THE WORKFORCE AND THE WORKING ENVIRONMENT

Over the past 200 years, the UK workforce has gone through many changes. In the nineteenth century, the Industrial Revolution resulted in many people, including women and children, moving from cottage industries to work in large factories.

Towards the end of the nineteenth century and at the start of the twentieth century, many factories started to lose work to overseas competitors and many people lost their jobs. The wars then caused major problems in Europe. Between the wars, Europe and the USA suffered major depressions where millions of people were out of work for long periods.

In the 50 years after the Second World War, many engineering and manufacturing companies improved production and, with a strong market, they were able to employ lots of people. However, as technology improved and production methods became more automated some factory workers found that their skills were becoming obsolete. Many workers retrained and gained new skills, but some found this difficult or even impossible.

Towards the end of the twentieth century, many engineering and manufacturing jobs had been replaced by machines carrying out automated processes. As automated systems improved the reliability and accuracy of production fewer workers were needed to service the machines.

Over recent years the UK has seen many companies move production overseas, where labour costs are lower, due to the lower wages paid to workers. These changes have meant that to be successful, engineers and manufacturing personnel have had to learn new skills. They have learnt to work in an industry that is ever-changing, where the machinery used is increasingly automated and controlled by computers. Companies now generally employ fewer, better-trained staff than in the past and the staff have to be more flexible in their approach to work.

Fig 3.20 Many companies have moved their production overseas to places like Asia

Some companies have altered their emphasis from making to designing. They come up with ideas and designs, and then make a prototype that can be used by an overseas factory to produce the product.

Throughout this period, the working environment has improved considerably. During the Industrial Revolution, factories were dirty, dangerous places.

Nowadays all workers in the UK are covered by the Health and Safety at Work Act, which makes it illegal to allow staff to work in a dangerous environment, or with complex equipment, without appropriate safety procedures in place. It also ensures that the health of staff is considered, meaning that they cannot be exposed to fumes and dirt that could be hazardous to health.

ACTIVITY

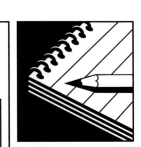

A company has been investigating different ways of manufacturing a product.

Method of manufacture	Manufacturing cost per unit	Outlay on equipment	Reject rate	Manufacturing time per unit
Manual	High	Low	High	Long
Fully automated	Low	Very high	Low	Short
Semi-automated	Medium	High	Medium	Medium
Outsourced	Medium	None	High	Very long

Using the information in the table and any other influences you can think of (such as volume of production), describe which of the manufacturing methods you would recommend to the company.

REVISION NOTES

- The pattern of employment has changed as technology has been introduced.

- Fewer people work in clerical jobs, as many of these are done quickly on computer.

- Many manual tasks have also been computerized.

- Skilled workers are still needed, but the skills they need are different.

- Most workers will have to retrain to keep their skills up to date.

⑦ IMPACT ON THE GLOBAL ENVIRONMENT AND SUSTAINABILITY

Engineering and manufacturing have enabled the Western world to achieve a higher standard of living, where items that would have been considered a luxury a few years ago are now common. The increased use of the world's resources that has fuelled this development has caused some people to become concerned over whether this standard of living can be maintained.

There are also concerns over the effects of the use of the resources on the planet itself. Pollution is blamed for changes in weather patterns. Chopping down rain forests, to retrieve natural resources, causes a loss of habitat for plants and animals. Soil erosion causes loss of agricultural land. Water-borne pollution from factory waste creates problems for marine life.

Fig 3.21 Deforestation can be devastating to natural habitats

Many new materials have been developed using relatively rare raw materials. Because these materials are rare, they are difficult to find and extract. This means that alternative materials or production methods are always being investigated. However as we become more reliant on new discoveries, alternatives can be hard to find.

Modern lifestyles also require huge amounts of fuel, much of which is still derived from fossil fuels. Many petrochemical experts believe that we have used up to 50% of oil reserves and that shortages will start to have an effect on prices over the next decade. There are large reserves of coal and gas, that we could use, but they create greater problems with pollution: coal releases a great deal more carbon into the atmosphere than oil. We could turn to nuclear or sustainable fuels like solar or wind, but at the moment they produce a small fraction of the energy we use.

There are also concerns over the rest of the world developing the standard of living we enjoy in this country. The world is a finite resource; we have already destroyed massive areas of natural habitat, both on land and at sea, to produce foodstuffs. With the world's population growing we will need greater amounts of food in the future.

As a nation develops its population consumes more. This is great news for engineers and manufacturers. In the long term, we will need to look carefully at what we produce and how we produce it, to achieve a sustainable future for everyone.

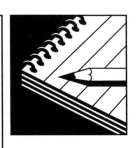

ACTIVITIES

1. Choose a product you have studied. Produce a diagram of its life cycle, from design to destruction.

2. Investigate your local recycling regulations and produce a help sheet describing how best to recycle/reuse or reclaim materials from the products you have studied.

REVISION NOTES

- Most people now accept that we need to protect our environment if future generations are to survive.

- Most people are less keen on having to do something about it if it costs them time or money.

- Manufacturers need to look at more efficient ways of producing and using energy and raw materials.

- We need to recycle more materials so that we can recover the ones we have already instead of making new ones, but this can sometimes use a lot of energy in the recycling process.

STAGES IN ENGINEERING AND MANUFACTURING A PRODUCT ⬡8

- Design
- Marketing
- Production planning
- Material supply and control
- Processing – production
- Assembly and finishing
- Packaging and dispatch

Earlier sections of this book have gone through the stages involved in designing and making a product in the engineering and manufacturing industry. For your GCSE you will carry out most of the stages yourself. However, in industry it is more likely that each stage of the design and manufacture of a product will be carried out by a specific individual or team.

Some companies use external agencies to help with their projects. They may commission market research to be carried out on their behalf by a specialist company, rather than employing staff in a marketing role.

It is relatively common for packaging materials to be 'bought in' from specialist suppliers. A design team may design the labelling and appearance of the package, but the package is often printed and prepared for use by an external company.

Wherever the stages take place, they have to be monitored and quality assurance and control systems have to be followed. A specialist department usually manages these systems, all checks carried out, and

any modifications made to the product should be recorded so that products developed in the future can incorporate any changes.

As discussed in previous sections each stage in the development of a product has been affected by the introduction of ICT.

ACTIVITY

1. Complete the table below stating two innovations that ICT has made to each stage.

Design	
Marketing	
Production planning	
Material supply and control	
Processing – production	
Assembly and finishing	
Packaging and dispatch	

---

⑨ INVESTIGATING PRODUCTS

An important aspect of any engineering or manufacturing project is to investigate what has gone before; it is never useful to reinvent the wheel!

Companies carry out exhaustive testing on products, both their own and those of their competitors. They tend to work through a sequence of investigations to test the product. In your investigations, you should try to do something under each of the following headings and questions.

- Research information from manufacturers and suppliers – take a look at web sites, publicity materials, and catalogues.

- Handle and examine individual products – make a physical examination of the items; take into account, size, quality of finish, and how appropriate it is to fulfil the task.

- Carry out simple tests to try to investigate the properties of the materials used, such as structure, heaviness, colour, texture, scratch and wear resistance, and areas that are most likely to be damaged.

- Evaluate the need for the technology, materials and components used. Do the materials make the best use of their properties?

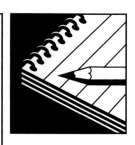

ACTIVITIES

1. Choose a product you have studied previously. For this product, complete the following.

   • Comment upon the role modern technology plays in the design and manufacture of the product.

   • Has the product, process or material replaced an older form of technology?

   • What are the benefits of using the technology to produce the item?

   • What are the implications of using new technology for the product and the manufacturer?

   • What is the purpose of the product? Does it satisfy a need?

   • Describe the structure and form of the product.

   • List and comment upon the materials and components used to make the product.

   • Describe, in detail, the technology used to design and make the product.

   In carrying out any investigation or test, you should write up a report detailing the following: title of test; apparatus used; aim of the test; expected outcomes; actual outcomes; conclusion.

2. Make a summary of what you found, for use in your examination.

Index

Note: page numbers in **bold** refer to key items.

INDEX

INDEX